SCIENCE, TECHNOLOGY AND DEVELOPMENT
North–South Co-operation

Edited by
MOZAMMEL HUQ
PRABHAKARA BHATT
CHRIS LEWIS
AHMED SHIBLI

Routledge
Taylor & Francis Group

LONDON AND NEW YORK

First published 1991 by Frank Cass and Co, Ltd

This edition published 2013 by Routledge
2 Park Square, Milton Park, Abingdon, Oxon OX14 4RN
711 Third Avenue, New York, NY 10017

Routledge is an imprint of the Taylor & Francis Group, an informa business

British Library Cataloguing in Publication Data

Science, technology and development: North-South
co-operation
I. Huq, Mozammel, *1940–*
338.926

ISBN 0 7146 3455 7

Library of Congress Cataloging-in-Publication Data

Science, technology and development : north–south co-operation /
edited by Mozammel Huq ... [et al.].

p.   cm.
ISBN 0-7146-3455-7
1. Science and state—Developing countries.   2. Technology and
state—Developing countries.   3. Science—Developing countries.
4. Technology—Developing countries.   5. Research—Developing
countries.   6. Technology transfer—Developing countries. Huq,
Mozammel
Q127.2.S43   1991
338.926'09172'4—dc20   91-36308
                                            CIP

Contents

cont.

Introduction: Science, Technology and Development: North–South Co-operation – An Overview

M. M. HUQ

This volume, containing the proceedings of a Conference organised by the Science, Technology and Development (STD) Forum in association with the Developing Countries Research Unit (DCRU), University of Strathclyde, on 9–10 April 1990, is a modest attempt at advancing awareness concerning Third World development mainly in two inter-related areas: (a) the problems faced in the advancement of science and technology in the Third World and (b) the use of scientific and technological knowledge for promoting sustainable development of these countries. Given the multi-disciplinary character of the Conference and, in particular, the variety of approaches adopted by the contributors, any attempt to synthesise the papers in a brief introduction is likely to be futile. The best that can be done is to introduce some of the papers in a brief review. But before that a few words on the objectives of the Conference, and in particular on the need for North–South Cooperation in science, technology and development, may be helpful.

The North–South Divide

The main themes addressed at the Conference were Technology Transfer from North to South, Scientific and Technical Manpower Development, the Current State of Science and Technology, and Indigenous Technological Capability.

The plight of the Third World countries, and especially the very poor (the World Bank's 'low income countries', with per capita incomes of less than $400 a year, and there are about 30 countries, including China and India, in this group; *World Development Report 1990),*[1] may be beyond comprehension by most of the people living in the North. What is particularly disturbing is the fact that some of these countries are becoming even poorer and that the proportional, as well as absolute, gap between the

Dr M. M. Huq is Senior Lecturer in Economics, University of Strathclyde, Glasgow.

1

rich and the very poor (excluding China and India) is widening, rather than narrowing.

An important reason for much of the South's inability to advance rapidly is its less than satisfactory rate of development of science and technology. Its pool of scientists and engineers is significantly lower in number, 195 per million compared with 2,792 per million in the North (3,233 in the USA, 4,836 in Japan and 5,414 in the USSR).[2] Its level of scientific research and development (R&D) is also very poor, the South spending 0.46 per cent of its GDP compared to 2.48 per cent in developed countries. The heavy dependence of the South on the North, often involving the almost uncritical import of machinery and equipment, is also not helping the cause of an effective transfer of technology to the South. It is, therefore, understandable that the countries of the South have strong feelings on the state of the existing relationship with the North, as is reflected in the recently published Report of the South Commission, *The Challenge to the South*.[3]

A decade ago, the *Brandt Commission Report* strongly advanced the case for North–South Cooperation,[4] but it appears that the appeal so persuasively advanced has failed to make the expected impact on many Northern governments. It is, however, wrong to say that people from the North would like to see their counterparts in the South remain in perpetual poverty and stagnation. On the contrary, there is a strong body of opinion in the North which earnestly wants rapid socio-economic development in the South. Our Glasgow Conference was organised with this spirit of North–South coooperation in mind.

Contents of the Volume: A Brief Review

The papers, reflecting in one way or another the theme of the Conference, are so diverse that it is difficult to classify them in groups. However, some of the papers are highlighted below in the order of their appearance in the text.

Anthony Clunies Ross discusses the need for North–South co-operation for environmental upgrading. This is a serious and significant attempt at establishing a solid international arrangement for pollution control and environmental improvement. The heavy polluters of the North have a major responsibility to subsidise the poorer nations in the South who are now facing a serious economic crisis, which is in part due to the environmental degradation caused by the industrialised North. Clunies Ross comes up with an environment protection formula:

Charging each 'overpolluting' nation so much per unit of pollutant emitted over a standard based on a target world per-head figure, and paying the proceeds to the rest (under-polluting nations) in proportion to the amounts by which they fall below the same standard, would give all an incentive to reduce, while preserving–equity and entailing large transfers to low-income countries.

Barker, Franceys and Pickford also take up the environment issue, but in a different, though related, context. While remaining committed to promoting health and quality of life in the South, they – as practising engineers – are of the opinion that the environmental upgrading can be achieved in proportional terms, at a lower level of specifications, compared to those in the North, because of what they call lower marginal affordability. They also argue that technology by itself will not lead to the necessary environmental upgrading, and they advance the case for user involvement at all stages – planning, implementation, maintenance and operation of technologies.

Indigenous technological capability, which is now attracting much interest in the technology transfer debate,[5] was viewed by a number of presenters during the Conference. James Pickett examined the issue in the context of post-war development in Sub-Saharan Africa. His paper, however, starts by viewing the standpoints taken by some of the great economists like Marshall and Schumpeter. According to Pickett, 'the main error of the post-independence period (in Sub-Saharan Africa) has been in the choice of strategy. The challenge of transforming economies in which much economic activity – including notably most of agriculture – is still in the household was underestimated.' He then goes on to add that 'the widespread failure of forced industrial development in Sub-Saharan Africa is not surprising. Domestic markets have remained small and the conditions for export competitiveness have not generally been met. . . . In these circumstances increasing industrialisation will normally require sustained improvement in agricultural productivity.' Such a standpoint, in Pickett's opinion, was obvious to Adam Smith, whose general framework of increasing division of labour supports growth from the frugal accumulation of capital and its application 'according to natural course of things'.

Helen Appleton and Andy Jeans take up the issue of technology transfer directly involving the people who are at the centre of the exercise. 'Technology transfer', according to them, 'is a process of sharing products, skills, and ideas that takes place over a period of time.' An important conclusion reached by the authors is that the inventiveness of human activity, especially at the grassroots level, needs to be recognised and that, in the design and development stages of technology transfer, both (poor) producers and consumers should be involved.

The need for technical manpower development for improving the ability of the Third World countries for an effective technology adaptation was the broad theme of a number of papers. Anders Närman takes up the issue with examples from three African countries. By viewing the modes of technical schooling/training in Kenya, Tanzania and Botswana, Närman argues for technical diversification and for more vocational training. Frank O'Reilly takes up the case of 'Farmer Education for Technology Transfer in the Third World', with supporting evidence from Northern Nigeria, Thailand and Libya.

Oraee and Haerian deal with technical manpower development in Iran and, in particular, the role of the present Iranian government in overcoming various constraints in this regard. In the current five-year plan of Iran, it has been proposed to increase the expenditure on education to 7.5 per cent of the GDP, a higher figure than in Japan, Canada and most other developed countries. According to this plan, there will also be a tremendous increase in trained scientific personnel, from the present level of 82 to 300 per million.

The effectiveness of technology transfer featured prominently in the Conference. In fact, George McRobie initiated the discussion by making a strong case for what has been commonly known as 'intermediate technology', technology that is easily grasped by users in the Third World. As he observes, 'the direct transplanting of rich-country technologies into the South has already done much damage to the interests of the poor.' He even goes further to add that 'the attempt to transplant Northern technologies into the South has been in general a disastrous failure'. Thus convinced, he has no hesitation in advancing the case for an appropriate technology which, according to his definition, 'will generally be small, relatively simple, inexpensive and (to be sustainable) non-violent towards people and the environment'. He goes on to add that

> to be appropriate the technology should be capable of local operation and maintenance, and local or at least indigenous manufacture; it should be owned and operated by its users, and result in a significant increase in their net (real or money) income; it should utilise to the maximum extent local or renewable raw materials and energy; and it should lend itself to widespread reproduction using indigenous resources and through the medium of local markets.

The above definition, by providing certain specific characteristics, attempts to make 'intermediate technology' appealing to a section of the people. Past failures on the part of many Third World recipients of large-scale Northern technologies have obviously made people like McRobie highly critical of such a development strategy of wholesale dependence on the North that completely ignores local capability.

The right approach will, therefore, appear to be to concentrate on advancing indigenous technological capability. To achieve this one does not go for self-sufficiency, but rather continues to adapt Northern technologies to suit local circumstances and also continues to advance the capability of the domestic machinery manufacturing sector. Most often what we observe, however, is that there is some form of transfer of 'production capability' from the North to the South, confined mainly to the transfer of machinery and equipment combined with some training as to how to use the hardware. Except in a few cases, there is no effective transfer of technology in the sense of 'investment capability' through the development of a local capital-goods sector with the participation of local engineers and scientists.

The paper by Huq and Islam, in the form of an industrial case study in Bangladesh, demonstrates the failure of *real* technology transfer because of an almost total uncritical dependence on imported technologies. In the process, Bangladesh has failed to take full advantage of the large investment which has gone into the development of, for example, the fertiliser industry in the country. However, this failure should not imply that Bangladesh can produce urea-fertiliser by using small, local technologies, because there is no such technology available. Indeed, the technology in urea fertiliser manufacturing has witnessed such an advancement, following R&D efforts in some multinational corporations from the North, that even the conventional large-scale reciprocating process has now been replaced with a modern variant with steam-driven centrifugal compressors, where the scale of production is significantly higher. In the process, the conventional technology has turned out to be obsolete.

The poor performance of small-scale technology by comparison with the large-scale variant was also demonstrated by Michael Tribe, this time in the context of dairy processing. Tribe considers the issue of technology choice and economies of scale by viewing the application of dairy processing technologies in the case of two dairy products – pasteurised milk and ultra-heat-treated (UHT) milk. Case study observations from Kenya show that there is really no place in the immediate future for a small-scale UHT plant.

Alistair Young's paper on the semiconductor industry raises both the weak and the strong sides of the transfer of the industry to a number of South-East Asian countries. It is a story of labour exploitation, as well as of skill development and employment creation. While not condoning the exploitative aspect of private foreign investment, Young finds grounds for support of such investment in the numbers who have found new employment as a result. Young is, therefore, able to conclude that to the Third World countries high-technology industry transfer 'offers mixed blessings, but they are blessings nevertheless'.

Hope for the South

Given that the issues involved are not only wide but at times complex, no attempt is made here at an extensive summary. Rather we list four main – but inter-related – points.

(1) The accumulated experience gained over the last few decades, in failures as well as in successes, can point the directions in which the Third World countries should proceed. For example, expenditures on R&D in the Third World need to be raised significantly.

(2) There has been a long debate over the last few decades on large-scale versus small-scale (intermediate) technologies, and in the recent past science-based 'high' technologies have emerged as a very important issue in the debate. Experience shows that they all have their own role to play in different contexts. What is essential is that developing countries need properly qualified manpower to assess competing technologies on their own merits and to select those that are found to be most suitable under the circumstances. In such an assessment, it is the *real* transfer, and not just the transfer of hardware, that should form the central focus.

(3) Since a genuine transfer of technology requires not only production capability but also investment capability, there is a need to emphasise the development of the capital goods sector in the Third World countries.

(4) With the concern now so strongly voiced in the North for environmental improvement, people can easily visualise the futility of viewing the North and the South as separate entities. Indeed, the time has come to quantify the responsibility of the developed North towards the undeveloped South, which has played an important role, even at the cost of environmental disasters, in the development of the North.

NOTES

The author is grateful to Professor A. Clunies Ross, and Drs P. Bhatt, A. Shibli and C. Lewis for their helpful comments.

1. See World Bank, *World Development Report 1990*, Washington, DC, 1990.

2. The figures refer to 1986, extracted from *UNESCO Statistical Yearbook* and shown in A. Salam, 'Science and Technology in the Third World', Paper presented at the Third General Conference of the Third World Academy of Sciences, Caracas, Oct. 1990.
3. See particularly Ch. 2, in *The Challenge to the South*, The Report of the South Commission (Oxford: Oxford University Press, 1990).
4. *North–South: A Programme for Survival*, The Report of the Independent Commission on International Development Issues under the Chairmanship of Willy Brandt (London: Pan Books, 1980).
5. See, for example, C. J. Dahlman, B. Ross-Larson and L. E. Westphal, 'Managing Technological Development: Lessons from the Newly Industrializing Countries', *World Development*, Vol. 15, No. 6 (June 1988). See also L. E. Westphal, Y. W. Rhee, L. Kim and A. H. Amsden, 'Exports of Technology by Newly-Industrializing Countries: Republic of Korea', *World Development*, Vol. 12 (May/June 1984), pp. 505–29; J. L. Enos and W. H. Park, *The Adoption and Diffusion of Imported Technology: The Case of Korea* (London: Croom Helm, 1988), and S. Lall, *Learning to Industrialize: The Acquisition of Technological Capability by India* (London: Macmillan, 1987).

ENVIRONMENTAL UPGRADING

Institutionalising the Costs of Environmental Upgrading

ANTHONY CLUNIES ROSS

An agreed international arrangement for providing governments with incentives for reducing certain kinds of environmental damage is needed, and might be pioneered in Europe. Charging each 'over-polluting' nation so much per unit of pollutant emitted over a standard based on a target world per-head figure, and paying the proceeds to the rest in proportion to the amounts by which they fall below the same standard, would give all an incentive to reduce, while preserving equity and entailing large transfers to low-income countries. A modified version of such a scheme seems practicable for discouraging carbon dioxide emission.

> When the Wind is fair and the planks of the vessel sound, we may safely trust everything to the management of professional Mariners; but in a Tempest and on board a crazy Bark, all must contribute their Quota of Exertion. The Stripling is not exempted from it by his Youth, nor the Passenger by his Inexperience. Even so in the present agitations of the public mind, every one ought to consider his intellectual faculties as in a *state of requisition.*
>
> S. T. Coleridge, 1795 [Holmes, 1989]

Those words were spoken shortly after the French Revolution. The present is another such time of 'agitations of the public mind', the most important for over 40 years, when events cause past assumptions to be abandoned. So much deemed politically impossible has occurred over the last two years that no sensible rearrangement of human affairs can any longer be deemed out of the question. It is as if we had once again the opportunity that was missed in 1919, and again in 1944–50, of proceeding to construct a world order of general protection and mutual responsibility.

The avenue has opened through the coincidence of the progress to

Professor A. Clunies Ross is in the Department of Economics, University of Strathclyde, Glasgow, Scotland.

11

Western European unity symbolised by the Single European Act and the maelstrom or vacuum created by the overthrow of tyranny in the former Soviet empire. Other events and discoveries in the 1980s have emerged to push us along the path of much enhanced common institutions.

One, if we care to consider it, is the shameful failure represented by the economic performance of most of the low-income primary-exporting countries. The second is the realisation that the risks of world environmental damage from activities now in full swing are much more serious than most informed people had guessed ten years ago.

The latter problem, of activities in any one place that contribute indiscriminately to worldwide environmental harm, is one of what economists call 'public goods', but transferred from the national to the international arena. Public goods are goods which are generally desired but which, because their benefits are not individually appropriated, may not be adequately provided unless the individuals who would benefit from them accept a superior power of enforcement and consequently taxation.

Measures to check ozone-layer depletion and global warming are in the short term costly. They may put any one country at a commercial disadvantage. Thus there will always be voices saying that such action is premature. The benefits of one nation's action in any case will not belong to that nation. They will as it were be thrown into the common pool and may seem small in the total world scene. Hence we need an international authority which will ensure that measures in one country that contribute to the world's flow of correctives are rewarded. The authority must have powers of enforcement. If it does not have them through commanding prevailing force, it must operate through the consent of those that do.

In the early 1970s when the environmentalists were commonly arguing for a zero-growth world economy, the discussion faced a political impasse. The poor nations naturally complained that zero-growth for them was much more costly than for the rich. Who would give way? Would the rich actually submit to a decline in income, and if so how would that be brought about? Thought, as exemplified by the famous Club of Rome study [*Meadows, 1972*], was impeded by the assumption of fixed coefficients. Growth, in population or income per head, was assumed to be matched by consistently related rises in energy and mineral use. History promptly supplied a corrective in the reactions to the great oil-price rises. It became clear that some of the critical coefficients certainly responded strongly to changes in relative prices. We do not need to abandon economic growth, which simply means the addition to, and increasingly efficient use of, the productive resources that we have. Rather growth has to be viewed in a fuller perspective.

Ideally we would say that the prices of resources and the penalties of pollution ought to reflect their true global long-term costs; but highly imperfect knowledge of what the effects of present action are prevents such a rule from being operational. Instead we probably need to accept national and international targets for particular dates and to apply price signals that give some prospect of reaching those targets.

How do we share the targets internationally? The temptingly simple approach is to say that everyone should cut output of CO_2 or whatever it is by say 20 per cent in 15 years. But this would rightly be thought unfair by those poor nations that contribute per head a tiny fraction of the amounts of most pollutants generated by the rich. Asking the same proportional reductions of Canada, which in 1985 consumed 9,200 units of commercial energy per head, East Germany, which consumed 5,700, and China, which consumed 500, would seem a very uneven form of evenhandedness. And a system that does not seem fair is simply unlikely to be accepted.

An alternative is to set targets only for the relatively rich nations, who after all produce most of the pollutants. The disadvantage of this is that it leaves no pressure on the now poor to economise on polluting activities. China may use only one eleventh as much commercial energy per head as East Germany, but it uses far more than other countries of comparable average income. And China's income may well grow fast, as over much of the 1980s.

There is another possibility, apparently very simple, which I shall try to explain in principle. Then I shall look at adjusting the numbers to make it workable, and project how we might move towards enforcing it.

I assume that measures to limit pollution will have actually to be applied by national governments, whether they work through pricing or through prohibitions and licensing. What is being discussed here is a system of international incentives to national governments so that they have, and know each other to have, a reason for imposing within their own jurisdictions measures for reducing the activities targeted.

In principle this could be done by setting international targets and fining governments whose jurisdictions fail to meet them. But 'fining' by a court of law seems inappropriate both symbolically and practically. Hydrocarbon use, for example, is not a crime. It is simply socially costly. It needs to be paid for over and above the cost of producing the material.

Instead I suggest a system of payments: from those countries that exceed standard amounts of pollutant to those that generate less. A tonne of CO_2 released into the atmosphere entails *prima facie* the same environmental cost whether it is released in Sweden or Swaziland. The principle might therefore be accepted (as a first attempt at solution) that, if there is a certain world target for CO_2 emissions from burning hydrocarbons for the year 2000, this should be allocated to the people of the world in equal amounts per head. At the end of the year, those countries that have emitted more than their allowance should pay into a fund a pre-agreed amount per tonne excess. If the fund is to be self-liquidating each year, the sum concerned should be divided among the governments of those countries that have emitted less than their quota, once again in proportion to the amounts below quota that they have emitted. As I shall suggest, it should probably not work exactly like this; nevertheless this is a simple, fair and rational starting-point from which modifications can be made.

The same principle could be applied to a number of pollutants, probably with the fund for each treated separately but with a single authority doing the arithmetic and determining the sums to be paid and received. The target for forests might be no net deforestation, or even a certain amount of negative net deforestation (positive net tree-planting) per head. An attempt might be made to have the defaulter payments for say CO_2 and nitrous oxide emissions roughly comparable in terms of their important adverse effects. Thus, if greenhouse effects alone are considered, a tonne of nitrous oxide would command 150 times the price of a tonne of CO_2 [*Boyle and Ardill, 1989: 28*], as in Sweden's emission taxes [*The Economist*, 1990]. Those that have several adverse effects might have a charge per unit made up of the sum of the charges for each effect.

By this means every country, whether high- or low-polluter, would have an incentive to reduce its polluting activity. India could expect to gain for each tonne reduction in CO2 or nitrous oxide emitted; but countries with low levels of pollutants per head would always be net recipients. The payments could be seen as going of right to the poor countries for putting less strain on the eco-system than the rest. They would naturally be earmarked either for measures actually to reduce polluting activities or for other forms of environmentally-friendly development.

Whether such a system as this could be acceptable to those governments likely to be net payers and also have the required incentive effects would depend on whether rates could be fixed small enough not to be an intolerable burden on the high polluters but still large enough per marginal unit to make it worth the while of most governments to respond.

Of all the pollutants of global importance, the really difficult one to deal with will very probably be CO_2. Any solution would probably stand or fall on its capacity to exert effective pressure for energy economy and for switching to non-hydrocarbon energy forms. For want of other figures that could be properly used to show the dimensions of a carbon tax, I consider this in very broad terms as if it were just a matter of cutting down commercial-energy consumption, and I use the world commercial-energy-use figures for 1985 from the World Bank's 1987 *World Development Report.*

A principal difficulty for a scheme such as I have set out is due to the enormous gap in typical energy use per head between poor and rich countries (especially some with very high use per unit income). Assume that after 1985 figures were known a decision had been made to set a global target for commercial-energy use for 1994, say, that represented a 10 per cent reduction over the nine years. The target would be 1,350 kg.-oil-equivalent units by 1994. Suppose that Canada (the extreme case) in fact made the 10 per cent cut. It would still be using 8,302 units a head, 6,952 units above the target average. If the population of Canada were unchanged, this would represent a total excess of 177 billion units. If the amount to be paid over by Canada were to be fixed by this excess but were to be held within one per cent of Canada's 1985 GNP (that is within $US3.5 billion), the amount charged per unit excess could be not more than two US cents. If all the high-polluters took part as well as all the low-polluters and all in fact reduced energy consumption by ten per cent, this would provide very considerable cash flows to low-income-low-polluters: $18.5 billion in the year to China, $18.0 billion to India, $184 million to Malawi (5.7, 8.7 and 15.5 per cent respectively of their GNPs). This charge per unit, however, would be only about 16 per cent of the cost of one kg. of crude oil at $US18 per barrel, a smaller proportion of course of the cost of 1 kg. of refined oil. Given the elasticities that have been estimated for the private responses of demand for energy products to price, such a proportion seems likely to be small in relation to the costs of conserving and switching. Even so, the charge for Poland (another extreme case) would be 1.7 per cent of its 1985 GNP on the same suppositions. Maybe the maximum scale of transfers that could be accepted might be more of the order of a one-cent-per-kg.-oil-equivalent charge, a very small proportional addition to costs.

So instead I suggest a compromise. The simple system set out above supposes that the 'standard' against which Canada's actual performance is judged is the consumption that it would undertake if it consumed at the world per capita average rate represented by the world target for the year. Because the gap between its performance and this standard is likely to be so large, the charge per unit for the gap must be small. But suppose that the standard for Canada were set not at the target world per capita rate but much closer to where Canada was actually likely to be. Then the charge per unit could be set much higher, say at ten cents, without involving Canada (or Poland) in unacceptably high total payments.

In another paper [*Clunies Ross, 1990*] I set out the details of such an alternative method (with its limitations and suggestions for taking some of the details out of political control) and also compare the device suggested here with an international system of tradable permits, as proposed, with the same objectives, by Michael Grubb [1989].

A 10-cent-per-marginal-unit charge on governments would be about twice as high as the most ambitious national attempt at a greenhouse tax. Sweden's new carbon tax [*The Economist, 1990*] (seven times as high as Finland's, the other main example) is about four US cents per kg. apparently of carbon (averaging something like five cents per kg. of oil-equivalent).

Could we see the nations binding themselves in this way for the common good? Certainly anything even vaguely like the arrangements outlined here would be very favourable to all the poorish nations. But would we, who take so much more from the world, who strain its tolerance so much more, enter into such a contract to encourage both them and ourselves, making net payments to them that have the attraction of being given and received by right? It would at least give us some earnest of each other's good behaviour, but it would do so not through international regulations and prohibitions, with legal penalties, which would inevitably, in such an area as this, be resented and have their absurd and inefficient by-products, but by encouraging national governments to take the least costly means available to induce the necessary changes.

If we in Europe see the sense of this kind of international regime for averting environmental disaster, then we could start the process by forming an 'environmental community' constitutionally modelled on the European Community but prepared to admit any nation that agrees to observe its rules. The device of Council of Ministers, Commission, Parliament and Court, has turned out to be an ingenious one for easing the process of pooling sovereignty. The relative powers of the institutions could be gradually changed as greater trust developed. If all the members of the existing Community would agree to form such an environmental community, membership of the latter might be made a condition of membership of the former. Since EC membership is now highly desirable, this would be a way of inducing other high-polluting powers to join. But membership should not be confined to those in the EC. EC success should not tie us down to its boundaries where they are not appropriate.

It should have its equivalent of the Commission, to which the Commission's environmental activities might be transferred. As it became wider in membership, it might also accept responsibility for dealing with environmental disasters among members, which assumes, of course, a taxing power. It should have an environmental parliament, probably with the members of the EC Parliament for the states that are also members of the EC acting as representatives of those states also in the environmental parliament. As the diversity of member-states and powers of the parliament increased, it might be necessary to have forms of qualified voting in the parliament, as well as (EC-style) in the council of ministers. The aim would be to ensure that the rules, the targets, and the rates of transfer payment, should be broadly acceptable to all parties, while discouraging individual member-states from blocking progress through special pleading.

REFERENCES

Boyle, S. and J. Ardill, 1989, *The Greenhouse Effect*, London: Hodder & Stoughton.
Clunies Ross, A., 1990, 'Transfers versus Licences as Incentives to Governments for Environ-mental Correctives', Paper for Development Studies Association Conference, Glasgow.
The Economist, 17 March 1990, pp. 66–7.
Holmes, R., 1989, *Coleridge: Early Visions*, London: Hodder & Stoughton.
Grubb, M., 1989, *The Greenhouse Effect: Negotiating Targets*, London: Royal Institute of International Affairs.
Meadows, D., *et al.*, 1972, *The Limits to Growth*, London: Pan Books.
Pearson, M. and S. Smith, 1990, *Taxation and Environmental Policy: Some Initial Evidence*, Institute for Fiscal Studies Commentary No. 19, Jan.
World Bank, *World Development Report 1987*, Washington, DC: World Bank.

Environmental Upgrading for Low-Income Communities of the South

PETER BARKER, RICHARD FRANCEYS AND JOHN PICKFORD

The experiences of the water supply and sanitation decade have led to a much better understanding of the range of technologies available for environmental upgrading of low-income communities in the South. Many technical problems have been overcome as equipment has improved and new materials have been developed such as plastics in pipes and handpumps and ferrocement in water tanks. Cairncross [1989] comments that 'a few general technical problems remain' but also that 'a far greater effort is required in the second half of the dual activity which industry calls R&D'. For it has become clear that the technology by itself will not lead to the necessary environmental upgrading. The technology alone is not sufficient to meet the objectives; that is, to promote health and quality of life. Unless the consumers are involved, systems will not be maintained correctly and will fail. There is therefore now a necessarily greater emphasis on community management, ensuring that the end users are involved at all stages of the planning, implementation, operation and maintenance of their own schemes.

Following on from this commitment to involve the beneficiaries there has been a new emphasis on cost recovery and financial viability [Briscoe and de Ferranti, 1988]. Unless the chosen system is affordable (irrespective of who actually pays for it) it is unlikely that sufficient funds will be available locally for maintenance. It is clear now that for a successful environmental upgrading scheme the different but interrelated components of health enhancement, social acceptability, technology, economic and financial viability, institutional support and environmental awareness all need to have been considered. This takes considerable skill, inevitably a higher level of skills per person served than much larger conventional systems. And for many projects these skills are required in

Peter Barker, Richard Franceys and John Pickford are in the Water, Engineering and Development Centre (WEDC) at Loughborough University of Technology, England.

the rural and peri-urban areas where the majority of the people still live but where the majority of the professionals are loath to live or work. Better design and implementation of a myriad small projects takes increased time and commitment.

Urban upgrading

All the statistics show that there is still a very large amount of work to do to meet the needs of the poorest, in addition to recognising the tendency for schemes to go out of action faster than new ones are installed. Considering firstly the peri-urban requirements, the most recent figures for the 'Low Income' and 'Middle Income Countries' [WDR, 1989] suggest that there is an urban population in the region of 1,439 million. Of these, best estimates indicate that at least 30 per cent – that is, 432 million people – are living in informal housing – that is, housing not sanctioned or approved according to government standards.

Using an arbitrary 20-year design horizon, the urban population in 2007 will be 3,695 million based on present long-term growth rates. Therefore housing and infrastructure will be required for an additional 2,690 million people (equivalent to 385 million houses) if we are to achieve housing for all in the urban areas within 20 years.

According to conventional standards, with household water supply, sewerage, solid waste collection, drainage and surfaced roads, the total annual cost per household (TACH) is in the region of $133 for these services.

The Affordability Gap

Considering the financial implications of a conventional solution for services, the total annual income per household for the poorest 40 per cent of the population in the 'Low Income Countries' is approximately $700. With affordability normally estimated in the region of 20 per cent for housing and services the total available annual expenditure is $140 (it should be noted that this is a generous affordability percentage, as studies show that the poorest can afford to spend least, even as a percentage).

Taking an assumed minimum standard 30m^2 house at a minimum cost for a semi-permanent building of $30/m^2 with 20-year repayment of a five per cent loan plus land gives a total annual cost per household of $100 for housing. Adding this figure to the $133 required for conventional services indicates an amount of $233, which is substantially in excess of presumed affordability of $140.

The normal solution of cross-subsidisation within a country is difficult where income distribution figures indicate that the poorest 80 per cent of Low Income Countries earn only $1,205 per household per year; the differential between the poorest 40 per cent on $700 and the poorest 80 per cent is not sufficient to provide a tax base to redistribute to the poorest 40 per cent.

Willingness to Pay

It is therefore necessary to reconsider the objectives of infrastructure. These objectives may be considered to be an attempt, primarily through the built environment, to obtain health benefits, security and social requirement benefits and lastly convenience benefits.

Housing and infrastructure standards reflect differing costs, risks and benefits. However, there is confusion between objectives and means; whilst health benefits are often used to justify investments, there is little evidence to suggest 'what benefit' accrues from 'what investment'. As time passes, and professionals learn by disasters ('to engineer is human'), there is a tendency to 'ratchet up' standards. That is, the standards or bylaws creep up over time but are never allowed to slip down again, just as a ratchet prevents a wheel or a bar from moving backwards. However, this does not necessarily mean that previous 'lower' standards are 'wrong', they simply carry a higher risk in health and safety terms; it should be stressed that any existing 'conventional' standards, for all their high cost, are not risk free. Therefore, as there is a cost/benefit ratio, it would seem reasonable to consider the use of a health risk/cost ratio. The means of infrastructure provision could then be more consistently related to the objectives.

Following on from this concept, it is necessary to consider alternative means for a standard of infrastructure provision which is believed to meet a minimum acceptable level of environmental health security. These ideas are illustrated in Figure 1. At a certain time a community's willingness and ability to pay for a standard of service is expressed by its demand curve. It is useful in this context to think of this as the community's Marginal Affordability curve. The curve records how much the community is able and willing to pay, given its income level, for alternative

FIGURE 1

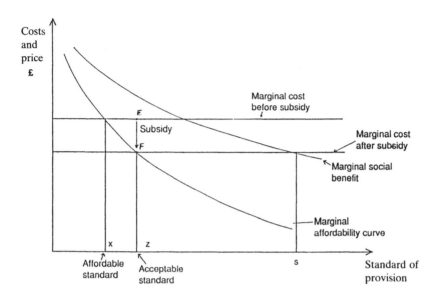

standards of service. It shows that demand for standard of service is constrained by two factors. The first is low per capita incomes and the second is ignorance of the social benefits conferred on others by a higher standard of service. Frequently the value of these social or external benefits is not perceived by consumers. Thus the benefits which accrue to others as a result of, say, a higher standard of housing – a lower probability of fire damage or the reduced possibility of disease spreading – are either not recognised by consumers or else are considered too much of a luxury to pay for.

Assuming the incremental or marginal cost of providing higher levels of service is constant, the intersection of the curves representing the community's Marginal Affordability and Marginal Social Benefit determines the current affordable standard of provision (X). Given the income level and perceptions of the benefit of service level, it may well be that the current affordable standard falls below the current acceptable level (Z). 'Acceptable' could be defined as sufficient to capture the key benefits of increased health and safety at costs commensurate with the community and agency resources. For this acceptable level to be attained Marginal Cost must be reduced by subsidy in the form of material or finance.

Alternatively, a subsidy must be paid to consumers, thus pushing the affordability curve to the right. Financial constraints and the desirability of maximising community participation suggest the former to be more practical.

By contrast, the Marginal Social Benefit curve incorporates the value of the external effects of higher standards of provision. Welfare econoics instructs that the socially optimal level of provision is at a standard (S). At lower standards of service the marginal social benefit is greater than the marginal cost of providing a higher standard of service, and so community welfare increases up to level S. This level of service is only attainable by a combination of subsidy, increased perception of the external benefits of higher standards and higher incomes.

In time the socially optimal level may become attainable, but it is argued here that more modest service levels (Z) may well capture the major health benefits. The convenience and comfort benefits of higher standards levels may only be reaped at undue cost and such standards ought to wait until incomes and perceptions of benefit permit.

The required level of subsidy to achieve the acceptable standard would be 'EF'; the achieved standard would be less than the social optimum but would reach the current limit of community understanding. To go beyond this level would represent wasted investment.

Any development scheme should thus be designed to allow communities to upgrade their levels of service as their understanding and incomes improve up to a maximum level of S.

Using a computer model to incorporate lifecycle costs of all the necessary services, relating services, land take along, with the interaction between different components, the cost of the simplest level of services provision may be calculated at a total annual cost per household of just $43. This is deemed to be affordable whilst ensuring environmental upgrading and has the potential for upgrading as the household desires and is willing to pay.

This low cost can be achieved by minimising access widths so that emergency vehicle access is possible at low speeds only. Storm drains are unlined and there is only one standpost for water supply in each cluster of houses. Solid waste is removed from a communal collection bin and there is no power supply. Perhaps the most significant change of standards is achieved by the use of on-site sanitation, even on plot sizes as low as $50m^2$. This removes the need for household water connections which a water borne sewerage system demands, as well as saving considerable amounts on buried sewer lines and the resultant operation and maintenance.

By these means health benefits are obtained at a level which is sustainable but which does not preclude upgrading to achieve convenience benefits as financial circumstances allow.

Rural Upgrading

Rural needs for environmental upgrading centre again on water supply and sanitation. Hundreds of millions of people still require improved facilities; in the main part, these will be household or community based, not suitable for utility-based distribution of networked services.

The very large number of discrete projects requires a different approach to service delivery. Two approaches are considered. The first is to accept that economies of scale do not always produce the desired benefits. Many systems could be reduced in size to more manageable elements which could then be *sold* to households and communities much like any other consumer item [*Franceys, 1988*]. A second approach is to enable engineers and other professionals to be more productive by use of the new information technology – that is, computers.

The use of computers, where they are seen as an everyday tool such as a calculator, can prevent mistakes, has the potential for improving the quality of the work and can free time for better control of construction quality or even for community discussions. There may be an added advantage in that the use of new technology such as computers is seen to confer status on their users and so enhances the level of professionals involved in low cost *appropriate* technology.

Computers can be used for the ordinary tasks such as word processing and more specialised tasks with spreadsheets to improve planning and financial information. They may also be used for specialist design techniques such as the preparation of water distribution networks. However, the main consideration in this paper is the use of *Knowledge Based Systems* (KBS) or *Expert Systems* which can assist the planner or engineer in solving a particular problem such as in the choice of a water supply or sanitation system.

Instead of textbooks a microcomputer is used and information is assessed in consultation between computer user and computer. Like a human expert, a KBS gives advice by drawing upon its own store of knowledge and by requesting information specific to the problem at hand from the user. The essential criterion for determining whether or not a subject is suitable for an expert system is the answer to the question 'can the knowledge on the subject be translated into "if-then" rules'; in other words, any rules on the subject should be encodeable in the format 'if a= b then c=d' [*Sawyer, 1984*].

Because of the limitations of adapting any rules to real life, computer

FIGURE 2
SAMPLE SCREENS FROM KNOWLEDGE BASED SYSTEM FOR SANITATION SELECTION

```
               Consideration of ANAL CLEANSING.

               Options available in this criterion are:
                  - water
                  - soft paper
                  - hard material
                  - bulky material

         Press the ENTER key to display comments on this criterion
         which would help you to decide on which option to choose.
 COMMENTS ON ANAL CL
```

Enter to select   END to complete   /Q to Quit   ? for Unknown

```
          Consideration of WATER available for flushing and/or anal cleansing.

               Options available in this criterion are:
                      -10 litres
                      - 3 litres
                      - 1 litre
                      - 0 litres

         Press the ENTER key to display comments on this criterion
         which would help you to decide on which option to choose.
 COMMENTS ON WATER
```

Enter to select   END to complete   /Q to Quit   ? for Unknown

```
         COMMENTS ON WATER AVAILABLE:

         The water supply service is an important criterion for selecting
         a sanitation system for an area.For households with only hand-carried
         supplies,conventional sewerage or septic tanks with soakaways are
         technically infeasible as are sewered pour flush systems since
         insufficient sullage would be generated (minimum recommended amount of
         sullage for successful running of such systems should be 50lcd).
         In households with piped supplies,feasible alternatives are narrowed
         down to systems which incorporate means of disposing of sullage such
         as sewerage,septic tanks,cesspits,aqua-privy and pour flush latrines.

             (Press any key to continue with consultation)
```

```
ZDDDDDDDDDDDDDDDDDDDDDDDDDDDDDDDDDDDDDDDDDDDDDDDDDDDDDDDDDDDDDDDDDDDDDDDDDDDDDL
.S                                                                          .S
.S               How much water is available for flushing                   .S
.S               and/or anal cleansing?                                     .S
.S                                                                          .S
.S       Note:                                                              .S
.S       If anal cleansing material is water,then the answer               .
.S       to this question cannot be 0 litres.                              .S
.S 10 LITRES               3 LITRES                      1 LITRE            .S
.S 0 LITRES                                                                .S
.S                                                                         .S
.S                                                                         .S
.S                                                                         .S
.S                                                                         .S
.S                                                                         .S
.S                                                                         .S
.S                                                                         .S
.S                                                                         .S
.S                                                                         .S
.S                                                                         .S
.S                                                                         .S
@DDDDDDDDDDDDDDDDDDDDDDDDDDDDDDDDDDDDDDDDDDDDDDDDDDDDDDDDDDDDDDDDDDDDDDDDDDDDDDDY
     Enter to select   END to complete      /Q to Quit    ? for Unknown
```

```
ZDDDDDDDDDDDDDDDDDDDDDDDDDDDDDDDDDDDDDDDDDDDDDDDDDDDDDDDDDDDDDDDDDDDDDDDDDDDDDD?
.S                                                                          .S
.S               The most appropriate method/methods of excreta-disposal    .S
.S               for the conditions given is a system of                    .S
.S               DIRECT SINGLE PIT POUR FLUSH LATRINES.                      .S
.S       NOTE:                                                              .S
.S       IF GROUNDWATER OR HARD ROCK IS ENCOUNTERED LESS THAN 2m BELOW SURFACE, .S
.S       THEN PITS MAY BE RAISED ABOVE GROUND LEVEL TO SUIT CONDITIONS.      .S
.S                                                                          .S
.SThe above named sanitation system(s) was/were based upon these criteria:  .S
.S                      ANAL CLEANSING MATERIAL:   water                     .S
.S                          WATER AVAILABLE:   3 LITRES                      .S
.S     AFFORDABILITY OF CAPITAL AND MAINTENANCE COSTS:   LOW                 .S
.S               POPULATION DENSITY OF PROJECT AREA:                        .S
.S                      DEMAND FOR FAECAL WASTE:   NO                        .S
.S            AVAILABILITY OF MECHANICAL PIT EMPTIER:   YES                  .S
.S         AVAILABILITY OF LAND FOR NEW/EXTRA LARGE PITS:                    .S
.S            PERMEABILITY OF SOIL IN PROJECT AREA:   PERMEABLE              .S
.S                                                                          .S
.S            (PRESS ANY KEY TO CONTINUE)                                   .S
.S                                                                          .S
@DDDDDDDDDDDDDDDDDDDDDDDDDDDDDDDDDDDDDDDDDDDDDDDDDDDDDDDDDDDDDDDDDDDDDDDDDDDDDDDY
```

systems are mostly used as human assistants (with humans always making the final decisions) rather than as 'stand alone' autonomous systems [*Gevarter, 1984*]. They are not there to replace a human expert, but to extend the experience and confidence of that expert.

The Knowledge Based Systems described below, developed at the Water, Engineering and Development Centre, represent a first stage of understanding of the knowledge required to improve water supply and sanitation programmes. In this respect they may be seen to convey 'shallow or surface knowledge' rather than the 'deep knowledge' which may be seen as the ultimate goal. However, it is the developer's belief that KBS's have a valuable role to play and as their effectiveness becomes apparent they can be enhanced.

A Knowledge Based System can run on a PC microcomputer. These machines are now counted in tens of millions and are appearing in government offices in all countries of the South. The program consists of two basic components: the 'knowledge base' and the 'inference engine'. The knowledge base contains all the rules and statements which represent the expert's decision-making logic, and the inference engine is the 'mechanism' which searches it to extract conclusions; that is, the process of developing evidence in order to arrive at new conclusions [*Hochgrebe, 1988*].

The major strength of a KBS over conventional computer programs is the ability to allow use to be made of qualitative knowledge (expertise and experience or 'soft' data), natural language and to allow interactive, user-friendly consultations. They are normally used to model complex situations with too many variables to be solved using a linear approach.

The relevance of KBS's for developing countries lies in their potential for high-level information transfer between local and international knowledge. This North-South co-operation and transfer of technology is particularly relevant where inter-disciplinary expertise is required.

Knowledge Based Systems for Rural Environmental Upgrading

Research at the Water, Engineering and Development Centre has led to the initial development of four prototype Knowledge Based Systems to assist community water supply and sanitation: 'Handpump selection' and 'Sanitation selection', described below, 'Water supplies in disaster relief' [*Ockelford, 1989*] to assist in emergency or refugee situations, and 'Design of small dams' [*Nichol, 1990*].

The KBS on handpump selection uses the experience gained in the UNDP/World Bank project on laboratory and field testing of handpumps

[*Arlosoroff et al., 1987*]. Because of the way in which the data were available, this particular program [*Hochgrebe, 1988*] acts substantially as the interactive front end of a database. Information on each pump is held on disk and the user is asked a series of questions regarding the maximum lift, maximum daily output, maintenance systems, corrosion resistance, abrasion resistance and manufacturing locale. At each step there is the possibility of choosing comments to explain the question and the reasoning behind answering it. As the user becomes more experienced these comments can be omitted. On conclusion of the questioning, the program searches the database for the pump or pumps which best fit the chosen criteria. The list is displayed and the user is offered further information on each pump, including the results of the testing programme, the country of production and address of supplier and an indicative price.

Hochgrebe concludes 'anyone who knows the particular project conditions for which a handpump is sought, but is not yet familiar with the selection approaches used, can obtain sensible results within 20 to 30 minutes by using the KBS which would be difficult to achieve using the original selection tables'.

The KBS on sanitation selection (Figure 2) was prepared by Eugene Larbi, using an algorithm prepared by Franceys [*1990*]. The program takes as its starting point the method of anal cleansing and offers comments as to why this is significant. Depending on the answer, the user is then asked about water availability, affordability, population density, demand for re-use of faecal waste, availability of mechanical pit emptier, land availability, and permeability of ground. Again, comments are offered at each stage but questions are only asked where they remain relevant. The program ends with a recommended sanitation system and a disclaimer to stress that the KBS is only a guide to the decision-making process and that the final choice still depends largely on the preference of the community or household.

Larbi [*1989*] believes that the program could be extended to include construction details of the chosen system, operational and maintenance requirements and typical costs and potential for householder involvement in construction.

Conclusion

Can the use of computers ever be justified for the promotion of low-cost water supply and sanitation? Or are they simply yet another inappropriate technology which will fail after an expensive trial run? The authors believe that knowledge-based systems have considerable potential to assist in overcoming the knowledge and skills gap which is limiting the replication of known technologies to millions of people.

Every possible method must be used to enable communities to take control of their own environment and to achieve the desired benefits. It appears that in the long term, computer-based information technology will play an ever increasing role in the storage and transfer of skills, knowledge and information. To meet the urgent needs of the millions without adequate water supply and sanitation it would be beneficial to use modern methods to enhance appropriate technology now – rather than allow these programmes to fall even further behind.

REFERENCES

Arlosoroff, S. *et al.*, 1987, 'Community Water Supply: The Handpump Option', Washington, DC: World Bank.

Briscoe J. and de Ferranti D. 1988, 'Water for Rural Communities, Helping People Help Themselves', Washington, DC: World Bank.

Cairncross S. 1989, 'Water Supply and Sanitation: An Agenda for Research', *Journal of Tropical Medicine and Hygiene,* 92, 301–14.

Franceys R. W. A. 1988, 'Social Impact of Appropriate Development on Village Life', Paper given to the Appropriate Development Panel, Engineering Management Group, Institution of Civil Engineers, London.

Franceys R. W. A., 1990, 'A Guide to Sanitation Selection', *Waterlines,* Vol. 8, No. 3.

Gevarter, W. B. 1984, 'Artificial Intelligence, Expert System and Computer Vision and Natural Language Processing', Park Ridge, NJ: Noyes Publications.

Hochgrebe E., 1988, 'Expert System on Handpump Selection', unpublished M.Sc. thesis, Water, Engineering and Development Centre, Loughborough University of Technology.

Larbi E., 1989, 'Expert Systems on Sanitation Selection', unpublished M.Sc. thesis, Water, Engineering and Development Centre, Loughborough University of Technology.

Nichol E., 1990, 'Application of Knowledge Based Systems to the Design of Small Dams', unpublished, Water, Engineering and Development Centre, Loughborough University of Technology.

Ockelford J. J., 1989, 'An Expert System for Water Supplies in Disaster Relief', unpublished M.Sc. thesis, Water, Engineering and Development Centre, Loughborough University of Technology.

Sawyer B., 1984, *VP Expert Manual,* Berkeley, CA: Paperback Software.

World Development Report 1989, Washington, DC: World Bank.

INDIGENOUS TECHNOLOGY, RESEARCH AND ADAPTATION

Indigenous Technological Capability in Sub-Saharan Africa

JAMES PICKETT

In the light of the marked shift in policy that has followed economic stagnation or worse, the selective concern of this paper is with economic growth and technological capacity in sub-Saharan Africa. The paper is in three parts. The first deals with economic theory and technical change, and so provides a broad context; the second examines salient characteristics of the African economies and of their economic performance since independence; and the third is concerned with the link between these, economic growth and technological capacity.

Introduction

For some, received general equilibrium theory cannot adequately handle technical change, since it makes of this either a parameter or a residual.[1] This apparent weakness may, however, be less damaging than is often suggested. Alfred Marshall believed that 'capital consists in a great part of knowledge and organisation', and that 'knowledge is our most powerful engine of production; it enables us to subdue Nature and force her to satisfy our wants'.[2] And his whole argument was driven by the quest for economic growth, which could ensure that 'all should start in the world with a fair chance of leading a cultural life, free from the pains of poverty and the stagnating influences of mechanical toil'.[3] Moreover, Marshall knew that progress first became possible when 'in the eighteenth century England inaugurated the era of expensive implements', so that subsequently 'on whichever side we look we find that the progress and diffusion of knowledge are constantly leading to the adoption of new processes and new machinery which economise human effort . . . we are moving on at a rapid pace that grows quicker every year'.[4] Marshall was also aware that there were extra-economic dimensions to economic growth, but still thought that economic progress was largely an economic matter. Why,

Professor James Pickett is Director of the David Livingstone Institute, University of Strathclyde, Glasgow. The author wishes to thank Martin Brown for helpful comment, without saddling him with any responsibility for the argument.

then, given his motivation did he develop a theory that was so careless, as it were, of a most important means to his end?

Rigour and the scientific status of the theory provide a possible answer. An evolutionary and dynamic theory would have defined a logical framework. Marshall's theory thus dealt with what was manageable and impounded other variables – however important – within *ceteris paribus*. This view does not entirely convince, even when the still-evident difficulties in designing an evolutionary theory are acknowledged.[5] Marshall's thought possibly reflected his perception of modern economies and economics in a way that did allow for technical progress.

In explaining the youth and immaturity of economic science in 1890, Marshall noted that the conditions of industrial life were themselves of recent date. Thus

> the economic conditions of modern life, though more complex, are in many ways more definite than those of earlier times. Business is more clearly marked off from other concerns; the rights of individuals as against others and as against the community are more sharply defined; and above all the emancipation from custom, and the growth of free activity, of constant forethought and restless enterprise, have given a new precision and a new prominence to the causes that govern the relevant values of different things and different kinds of labour.[6]

And in rejecting competition *per se* as *the* distinguishing feature of industrial life, Marshall records that

> there is no one term that will express these characteristics adequately. They are . . . a certain independence and habit of choosing one's own course for oneself, a self-reliance; a deliberation and yet a promptness of choice and judgement, and a habit of forecasting the future and of shaping one's course with reference to distant aims. They may and often do cause people to compete with one another; but on the other hand they may tend . . . in the direction of co-operation and combination. . . . But these tendencies towards collective . . . action are quite different from those of earlier times, because they are the result not of custom, not of any passive drifting into associaton with one's neighbours, but of free choice by each individual of that line of conduct which after careful deliberation seems to him the best suited for attaining his ends, whether they are selfish or unselfish.
>
> The term 'competition' has gathered about it evil savour, and has come to imply a certain selfishness and indifference to the well-

being of others. Now it is true that there is less deliberate selfishness in early than in modern forms of industry; but there is also less deliberate unselfishness. It is deliberateness, and not selfishness, that is the characteristic of the modern age.[7]

This emphasis on free and deliberate individual choice that best serves individual purpose leads naturally to the exacting analysis of the behaviour of individual economic agents – consumers and firms – at the heart of Marshall's *Principles*. Yet free and deliberate individual choice are also at the heart of Marshall's view of economic progress. How then is this to be reconciled with the emphasis given to 'expensive implements'? Through, runs the desperately brief answer, the importance attached to industrial organisation, efficient resource allocation, and the surplus. As Schumpeter has reminded us, Marshall's affinity with Adam Smith was great.[8] Both certainly put much stress on industrial organisation, what Smith called the division of labour. Marshall recognises that the insight that labour efficiency increases with organisation is as old as Plato. 'But', he adds immediately, 'in this, as in other cases, Adam Smith gave a new and larger significance to an old doctrine by the philosophic thoroughness with which he explained it, and the practical knowledge with which he illustrated it.'[9] And if the weight Adam Smith placed on the division of labour is now legend, it should be noted that he saw it, *inter alia*, as the source of technical progress. Growing markets were as stimulating to invention and innovation as they were to investment. Thus, although in time dynamic gains in output may normally be expected to exceed those from a more efficient resource allocation than ever before, static efficiency and dynamic gain are not independent but sequential parts of a seamless process. And in a society that looks to allocative efficiency, technical progress will look after itself.

To argue thus is, of course, to oversimplify. One of the three duties that were left to the sovereign in Adam Smith's system of natural liberty was the provision of public goods; and, as is widely-known, Marshall saw clearly the challenge increasing returns offered to the efficiency of the market. Still, the above reasoning lessens the urgency of finding a more comprehensive and convincing theory. Moreover, it also reminds us that public action to promote technical progress is only justified if it clearly improves on spontaneity.

The adequacy of allocative efficiency has, of course, been challenged. Almost alone among major economists of his generation, Schumpeter placed technical change at the centre of the capitalist system. And he gave due weight to social and institutional factors and to the link between these and economic elements.

Schumpeter made the entrepreneur the principal – and a heroic – actor, and innovation the central act, in his attempts to unravel the dynamic of capitalist growth. The business game, he insisted, was not roulette, but poker. The size of the kitty, and the beguiling combination of skill and luck, were sufficient 'to attract the large majority of supernormal brains and to identify success with business success'. The talent thus attracted was not preoccupied with static resource allocation. Success did not lie down that road, but was to be found in dynamic innovation – not in doing old things better, but in doing new things differently. Business strategy is only to be understood in the context of 'the perennial gale of creative destruction'. Thus

> Economists . . . look . . . at the behaviour of an oligopolist industry – an industry which consists of a few big firms – and observe the well-known moves and countermoves within it that seem to aim at nothing but high prices and restrictions of output. They accept the data of the momentary situation as if there were no past or future to it and think that they have understood what there is to understand if they interpret the behaviour of those firms by means of the principle of maximising profits with reference to those data. In other words, the problem that is usually being visualised is how capitalism administers existing structures, whereas the relevant problem is how it creates and destroys them . . .
>
> As soon as quality competition and sales effort are admitted into the sacred precincts of theory, the price variable is ousted from its dominant position. However, it is still competition within a rigid pattern of invariant conditions, methods of production and forms of industrial organisation in particular, that practically monopolises attention. But in capitalist reality as distinguished from its textbook picture, it is not that kind of competition which counts but the competition from the new commodity, the new technology, the new source of supply, the new type of organisation (the largest-scale unit of control for instance) – competition which commands a decisive cost or quality advantage and which strikes not at the margins of the profits and the output of existing firms but at their foundations and their very lives. This kind of competition is as much more effective than the other as a bombardment is in comparison with forcing a door, and so much more important that it becomes a matter of comparative indifference whether competition in the ordinary sense functions more or less promptly; the powerful lever that in the long run expands output and brings down prices is in any case made of other stuff.[10]

Moreover not only was technical progress thus central to capitalist development, it was also discontinuous. As Schumpeter himself put it.

> The historic and irreversible change in the way of doing things we call 'innovation' and we define: innovations are changes in production functions which cannot be decomposed into infinitesimal steps. Add as many mail-coaches as you please, you will never get a railroad by so doing.[11]

Such language and thought do represent a powerful challenge to the importance of static resource allocation. Some nevertheless believe that 'Schumpeter is not enough'.[12] He neglected the Third World, paid insufficient attention to international trade and the international diffusion of technology, failed to formalise his models, and – for all of his insights into innovation, economic dynamism and the monopoly profits that could flow for a time from technical advance – remained too attached to Walrasian general equilibrium. There is therefore a need for a general theory that would rival Walras in rigour, but exceed him in coverage – because of better and more apt treatment of technical change. To go in this direction would place developing countries in an international setting, and look for improvement in international trade theory and in understanding of the 'catching-up' mechanism.[13]

Short of definitive resolution of the conflict between the Smith–Marshall and neo-Schumpeterian views, it may be noted that each may have merit, depending on the level of development. Schumpeter was concerned wth mature capitalism, which is hardly an apt description of the sub-Saharan economies. And if the neo-Schumpeterians would extend the system to embrace economies at different levels of development, their justification for doing so may rest on views on the importance and character of technology transfer that may not be well-grounded. What, run the important questions, are the salient features of the African economies? How have they performed? How can they now best grow?

Characteristics and Performance in Sub-Saharan Africa[14]

The sub-Saharan countries do not self-evidently constitute a homogeneous group. Yet for the most part they have comprehensive poverty in common, and this already well-known fact may be quickly confirmed from the World Bank data of Table 1. Thus, of the 34 African countries covered, 26 – well over two-thirds – were in the low-income category in 1987. Moreover, 11 of these were in the group of 15 countries worldwide in which income per head was US$250 or less. More generally, the higher the income category, the lower the African representation, so that there were no sub-Saharan countries in which average income in 1987 was more than US$6,000, and but one (Gabon) in which it was more than US$2,000. Reflecting low income levels, most sub-Saharan countries were economically small. Thus in 1987 the aggregate production of 20 of them was less than US$3,000 million, which was about eight per cent of Hong Kong's total output; and none of them had a gross domestic product in excess of US$25,000 million. There was, therefore, no sub-Saharan economy as big as that of Hong Kong.

Why are the sub-Saharan countries poor? One answer is the dominant weight of low-productivity agriculture. The relative importance of peasant agriculture is, however, not a cause, but a consequence of poverty. Because African countries are poor their ability to undertake investment in physical and human capital is limited; and because their markets are small the incentive and the capacity to innovate are also extremely weak. Being poor is a result of severe resource limitation, which requires that a large proportion of available resources should be applied to basic food production. Such application does not guarantee, however, that people will have enough to eat.

Though it cannot provide a causal explanation for poverty, economic structure is nevertheless important, since it is the only realistic starting point for economic transformation. More detailed characteristics and information on change between 1965 and 1987 are captured in Table 2. This starkly reveals just how woeful Africa's economic performance has been. Thus between 1965 and 1987 total output in sub-Saharan Africa as a whole barely kept pace with population increase, so that average income was little higher towards the end of the 1980s than it had been 20 years earlier. In the lower-income economies even this modest progress was not achieved since, whether or not Nigeria is included, income per head declined between 1965 and 1987. And the failure to grow was reflected in the data on economic structure. The distinctive pattern of change normally associated with economic progress – a fall in the relative standing of agriculture and a rise in that of industry – is not apparent in the figures of Table 2.

Why did sub-Saharan Africa do so badly? Many of the African economies were, of course, poorly prepared for modern economic growth as they came to independence in about 1960. Educational systems were small, infrastructure was limited, and specialisation had not gone very far. Nor were natural endowments generally impressive. Then, as indeed still, much of the African economy consisted of labour and land. And if the labour was largely unskilled, much of the land was difficult to farm and only grudgingly productive. Agriculture was, of course, rain-

TABLE 1

ECONOMIC STANDING OF SELECTED ECONOMIES, 1987

(a) Income Level

No of Countries

Income Group	Sub Saharan Africa	Asia	L.America and Caribbean	Total[a]	SSA Share of Total (per cent)
Lower-Low Income (Less than US $250/head)	11	4	0	15	73.3
Upper-Low Income (US $260-450)	15	5	1	22	68.2
Lower-Middle Income (US $451-2000)	7	3	14	34	20.6
Upper Middle Income (US $2000-6000)	1	1	6	15	6.7
High-Income (US $6000 +)	0	3	0	25	0.0

(b) Market Size

Gross Domestic Product

Less than:

		Sub Saharan Africa	Asia	L.America and Caribbean	Total	SSA Share
US $ 1,000	million	5	2	0	8	62.5
US $ 1,001-3,000	million	15	1	2	19	78.9
US $ 3,001-6,000	million	8	0	9	20	40.0
US $ 6,001-12,000	million	3	1	3	9	33.3
US $12,001-40,000	million	1	5	2	19	0.0
US $40,001-150,000	million	0	2	5	19	0.0
US $15001-1,000,000	million	0	2	1	13	0.0
US $1,000,001+		0	1	0	3	0.0

[a]Totals sometimes include non-Asian developed economies and Middle East Economies.
Source: World Bank, *World Development Report, 1989*, World Development Indicators, Table 1.

TABLE 2

CHARACTERISTICS OF AFRICAN AND OTHER ECONOMIES, 1965, 1987 AND 1965–87

	GNP per Head (US $, 1987)		Av. Annual Rate of Growth, 1965–1987, GNP per head (per cent)	Distribution of GDP							
				Agriculture		Industry		Manufacturing		Services	
	1965	1987		1965	1987	1965	1987	1965	1987	1965	1987
Sub-Saharan Africa	316	330	0.2	43	34	18	28	9	10	39	39
Low-income	273	270	-0.1	45	38	18	25	8	8	37	37
Low-income excluding Nigeria	253	240	-0.2	42	40	19	19	9	9	39	42
Non-African Low-Income	130	295	3.8	–	–	–	–	–	–	–	–
Excluding China and India	132	288	3.6	–	–	–	–	–	–	–	–
Middle-Income	1,050	1,810	2.5	–	–	–	–	–	–	–	–
High-Income	8,750	14,430	2.3	–	–	–	–	–	–	–	–

Source: *World Bank Report 1989*, 'Sub-Saharan Africa: From Crisis to Sustainable Growth' and *World Development*.

fed and so sensitive to weather conditions. Thus those who would excuse African economic performance can point to the above-average incidents of drought in the 1980s. They can also note the post-1973 deterioration in world economic performance, invoke the two oil-price shocks and the unexpectedly severe and prolonged 1980–1982 global recession, and make the most of fluctuations in Africa's terms of trade.

Yet great caution is needed. The comparative data of Table 2 make it clear that the poor African economies did much less well than other low-income countries. With or without the fast-growing Indian and Chinese economies, the non-African low-income countries recorded much higher rates of growth than did their African counterparts between 1965 and 1987. These were indeed the fastest growing group of countries in the world economy at the time. And if disaggregation among the middle- and high-income countries would cause that statement to be qualified, the fact remains that a number of Asian economies clearly out-performed the poor African ones, though the Asian countries also began with low income levels, limited resource endowment, and small market size.[15]

This makes it necessary to consider policy. And this, it may be stated baldly, was wrong-headed. Possibly under the influence of prevailing economic thought, policy generally favoured industrial development and a state that was economically very active. At independence, however, the African economies were certainly poor but not labour surplus. There was, therefore, no strong economic reason for favouring industry at the expense of agriculture. Nor was the economically active State ever likely to impress. A general point of long standing was made with unsurpassable incisiveness by Henry Sidgwick. 'It does not follow', he wrote, 'that whenever laissez faire falls short government interference is expedient; since the inevitable drawbacks of the latter may, in any particular case, be worse than the shortcomings of private enterprise.'[16] In Africa the economic prowess of the State was specifically and fatally weakened by lack of data, knowledge and skilled manpower. Even if central control and extensive State participation in the directly productive economy does have a place in the scheme of things, that place has not been the sub-Saharan Africa of the last 30 years or so. It may, of course, be objected that the constraints of State efficiency operate on private effort also. The data and knowledge required by the private operator are, however, local; and he or she is well able to judge how best to use whatever skills are available in this narrow context. For the State to run the economy its data, knowledge and skill requirements are economy-wide. It has to co-ordinate as well as participate. And in African – perhaps in any – conditions that is not efficiently possible.

Thus the main error of the post-independence period has been in the choice of strategy. The challenge of transforming economies in which much economic activity – including notably most of agriculture – is still in the household was underestimated. And the ability of nascent states to make good market failures was as much over-valued as the virtues of even imperfect markets were down played. Resource allocation was inefficient with consequences that, since they were cumulative, were disastrous. Partly because the needs of public expenditure were thought to be served by it, a grossly inflated exchange rate was at the heart of import-substituting industrialisation. But the items on the import list are evidence of demand, not indicators of efficient supply. Nevertheless countries acted as if they could compete, mainly by taking protective steps to weaken and sometimes to strangle the competition. This policy doubly discriminated against agriculture. It distorted resource allocation directly and it further weakened agriculture through its reliance on an unjustifiably high exchange rate. This cheapened food imports, and made it more difficult than it would have been otherwise to pay domestic farmers prices that would have encouraged increased production. Here is an important part of the explanation of Ghana's lost pre-eminence in the world cocoa market. Moreover, the exchange rate is a market price. If the foreign price of the domestic currency is artificially low there will never be enough foreign currency. Parallel markets and corruption thus flourish. And, as Adam Smith remarked of the monopolistic East India Company, one cannot blame people for taking self-interested advantage of circumstance. It is the circumstance – the institutional framework – that has to be altered.

What has all this to do with technical progress and economic theory? The flaunting of conventional theory, runs the reply, is in effect an important part of the explanation of the disappointing lack of economic progress. Neo-classical analysis does identify the main reasons for the slow growth. Though the two are linked, diagnosis can, of course, run ahead of prescription, so that it need not follow that standard theory does equally well in discerning the way ahead. Even then, however, the theory does better perhaps than many would think.

Economic Growth and Technology in sub-Saharan Africa

In sub-Saharan Africa the economic challenge lies still in the transformation of largely agrarian economies. In the low-income countries agricultural output even now accounts for upwards of 40 per cent of GDP, and the sector employs a much larger proportion of the labour force and supports an even higher proportion of the population. Agricultural products still dominate the export list. In the poorest countries more than 70 per cent of the labour force is engaged in food production at a low level of productivity. In the United States a farmer produces enough food for more than 100 people, but in Ethiopia it is often more than a farmer can do to feed his family and himself. In the poor African countries specialisation is extremely limited. And much of it is in the poor peasant household, so that the opportunities for market exchange are not great, and the integration of the domestic economy is still far from complete. Such industry as there is is generally inefficient, and no industry-led African economic miracle is in prospect. Indeed, modest increase in the export of industrial goods would be a big achievement.

Limited growth notwithstanding, contemporary sub-Saharan Africa is different in important respects from what it was 30 years ago. Population has grown and arable land may – because of degradation – have shrunk. The urban population has also grown, more quickly than the total. There has been at least the possibility of industrial learning-by-doing, and there certainly are more powerful urban vested interests than before. Still the lack of progress, the prospect of falling even further behind, and the thus congenial policy blandishments of the World Bank *et al.* have led to marked policy change, with more reliance on markets and the price mechanism being at the heart of policy reform. It is expected that more realistic exchange rates and other changes will benefit agriculture as market forces are allowed more scope than hitherto. In the meantime, sub-Saharan Africa has been operating well within its production frontier. Achieving efficiency would thus raise output and output per head. There are, however, limits to the growth that could result within the constraints of existing resources so that the real challenge is that of shifting the production frontier outward. And the central question is whether that task can be left wholly to the market. Innovation and investment (in human and physical capital) are called for; and though spontaneous development is possible, it has to be asked if apt and judicious State action could not accelerate progress.

The position taken here is that it can – provided that the lessons of experience are properly learned, that government acts in keeping with its comparative advantage, and that government works with rather than against the grain of the market. Where, the important question now is, does the comparative advantage of government lie? Not, runs the first answer of theory and experience, on the state farms or the public corporation. Production (and distribution) are best left to the market. Governments should do what others cannot or (given the private economic calculus) will not do in the public interest. Even this, however, is broader than public skills can handle, so that tighter priorities are necessary. These can be found in economic strategy, so that it is convenient to turn to this before considering technology.

Strategy

In retrospect, the widespread failure of forced industrial development in sub-Saharan Africa is not surprising. Domestic markets have remained small and the conditions for export competitiveness have not generally been met. Viable industrial development has thus been confined to products for which the minimum efficient plant size has been also small, and in the manufacture of which local provision has transport or other cost advantage over foreign. Thus the dynamic benefits – learning-by-doing, and scale economies, for example – that were to accompany import-substituting industrialisation were not realised. Indeed total factor productivity growth has often been negative.[17] The limited division of labour that has characterised the poor African societies has meant that the weight of the market in total activity has been small and that market experience has been limited in risk, scope and scale. In these circumstances increasing specialisation will normally require sustained improvement in agricultural productivity.

This would have seemed – indeed did seem – obvious to Adam Smith. Within the general framework of increasing division of labour, growth comes from the frugal accumulation of capital and its application, 'according to the natural course of things', such that it is

> first directed to agriculture, afterwards to manufactures, and last of all to foreign commerce. This order is so very natural, that in every society that had any territory, it has always . . . been in some degree observed. Some land must have been cultivated before any considerable towns could be established, and some sort of coarse industry of the manufacturing kind must have been carried on in those towns, before they could well think of employing themselves in foreign commerce.[18]

And before Adam Smith's famous 'natural progress opulence' is imprudently dismissed as being dated, it should be remembered that the economies of which he was writing are in important respects closer to those of contemporary Africa than those that interest Schumpeter. It is therefore not surprising that Sir Arthur Lewis should have echoed Adam Smith in his advice to the government of what was then the Gold Coast in 1953. It was, he urged, first necessary to develop agriculture, since the main obstacle to growth was stagnant productivity in that sector. Unless agricultural output per worker was increased it would not be possible to provide the market, capital and labour for industrial development. Nor would industrial development be efficiently possible without the requisite infra-structure. Lewis's advice was, of course, ignored, but the consequences of it not being taken were so awful that they now stand testimony to its continuing relevance.[19]

Thus, the question remains: how is agriculture to be developed? By, runs a plausible, experience-based, answer, leaving production and marketing in private hands, designing macroeconomic policy to maximise incentives for efficient production, and looking to the government for a lead on agricultural research and development, extension, and the provision of infrastructure. These conditions met, agriculture cannot merely facilitate economic growth, but lead it.[20] Moreover, drawing on Ricardo's analysis of the growth effects of the distribution of the gains from trade, Ronald Findlay has reminded us that African economies should be well-placed – because of their peasant structure – to plough back the fruits of their agricultural comparative advantage.[21] Equity as well as efficiency could thus be served. It remains, of course, to be shown that boosting agriculture would yield income growth with a helpful spread of income elasticities of demand. It is, however, plausible that as peasants became richer the scope for economically-profitable non-farm rural acti-vity would increase more than proportionately, and that the pattern of industrial growth would more gradually from the small- and medium-scale and often rural-based to the urban and large-scale. It is thus appro-priate to turn to technical change in this agriculturally-biased process.

Technical Change

It is easier to catalogue the gains that should come in principle from agricultural research and development and related activities than to achieve them in practice. This is largely because not enough is known about present agricultural practices, about the likely pay-off to alternative research strategies, and about the capacity of governments to design let alone implement a programme with an optimal allocation of effort between the private and the public sectors.

Still, there would be widespread agreement, which would include market-inclined economists, that government has played an important role historically in the development of agriculture, particularly in relation to research, extension, education and training – where externalities have been important. More recently in rapidly-growing countries such as Taiwan, a general reliance on the market has not prevented high levels of surplus extraction from the agricultural sector. Correspondingly high

levels of public investment in agricultural technology, irrigation, transportation and primary education, however, have ensured high rates of
growth of agricultural productivity. This confirmation of the importance
of public activity should not be overlooked in Africa, though recollection
should be matched by realistic awareness of government limitation.

Experience elsewhere also confirms that the scope for straightforward
technology transfer in agriculture is limited. The ability of poor countries
to take advantage of foreign and international advances and to borrow
specific elements of agricultural technologies from abroad depends on
their own scientific capability, as measured by the level of trained manpower and the effectiveness of their research organisations, so that these
things too have to be developed within the constraints that operate on
poor-country government. And in this regard many questions arise.
What proportion of total activity should be undertaken by the State?
Indeed how is this question to be answered? How much attention should
be paid to the eradication of the sources of disease that keep population
away from well-watered areas in much of Africa? Should this command
more resources than the search for faster-maturing higher-yielding seed
varieties, particularly since the only clear success in this respect to date
has been with maize in relatively fertile areas in East and Southern
Africa? Or should irrigation be centre stage? What moreover should be
done to reduce the risks that African peasants run from uncertain yield
and uncertain prices? To what extent would better roads and increased
and improved credit reduce these risks? And how does the extent and
character of the needed local knowledge and skills vary across technologies. Are bio-technology and information technology relatively un-
demanding in these respects, so that foreign inventiveness and rural
education could deliver much progress?

These important questions themselves dictate a lengthy and lively
research agenda. That answers are so far largely still to be found means
that present prescription has to be cautious. Some things are nevertheless
clear. The first is the importance of national research systems and the
linking of these to efficient extension services. Agriculture is above all a
location-specific activity, so that successful adaptation of, for example,
international breakthroughs in plant development is only possible on the
basis of adequate local knowledge. Hence the importance of a national
research network. It is also clear that rain-fed agriculture has been increasingly hazardous in Africa. This points to a growing importance for
irrigation. Experience with large-scale irrigation has not, however, been
encouraging so that much work remains to be done on suitable forms of
water delivery. In this and in other respects there is a great need for farm
studies.

It is, of course, not necessary to have the results of these to be convinced of the importance of giving public priority to agricultural research, extension and rural infrastructure. Location-specific insights are needed, however, if public policy is to be comprehensive and successful.

Concluding Comment

This paper has moved unevenly from general equilibrium theory to the need for a multitude of location-specific studies. It has recognised the limitations imposed by ignorance, and so been more general than desirable in discussing the desiderata of technical change. The paper has nevertheless focused on the major characteristics of the low-income African economies, and reasoned from these to economic and technology policy.

Without going much beyond Adam Smith and Marshall, the importance of the agricultural surplus has been emphasised. And, in accepting that this has to be increased, it has been suggested that production and distibution are best left largely to the market. It has also been suggested, however, that progress can be greater and quicker if the State gives priority to agricultural research and related activities, the pressing need being that of increasing agricultural productivity. Though there is a large and growing literature on agricultural research and development, this emphasis on the farm has not been a prominent feature of the general and larger literature on the choice, adaptation and transfer of technology. There industry has been at the centre of the analysis. From an African standpoint, however, that is definitely incongruous. Industrialisation remains the goal, even in sub-Saharan Africa. At the present stage of development, however, the important question relates to the availability of labour-intensive techniques.

It could therefore be argued that much of what has to be decided about agricultural and industrial technology falls within the rubric of the choice of technique. And if imaginative reading of Adam Smith and Alfred Marshall is a very adequate basis for general economic strategy, neo-classical theory is equally adequate for technology policy. It does not, of course, follow that an economic theory that could deal more robustly with technical change than neo-classical thinking would be unwelcome. It would, however, be a mite precious in relation to sub-Saharan Africa. There the question perhaps is not what is to replace equilibrium theory, but whether a sufficient number has experienced 'the emancipation from custom, and the growth of free activity, of constant forethought and restless enterprise' that Marshall placed at the heart of modern economic life.

NOTES

1. For a recent collection of essays on this theme, see G. Dosi *et al.* (eds.), *Technical Change and Economic Theory*, (London: Pinter, 1988).
2. A. Marshall *Principles of Economics* (London: Macmillan, 1890), 1920 edition, p. 115. The link between Marshall's analysis and general equilibrium theory is clear from his Note XXI, ibid.
3. Ibid, p. 3.
4. Ibid, pp. 184–5.
5. See, for example, Freeman's introduction to Dosi, op.cit.
6. Marshall, op.cit., p. 4.
7. Ibid, pp. 4–5.
8. J. A. Schumpeter, *History of Economic Analysis* (Oxford: Oxford University Press, 1956), pp. 833–9.
9. Marshall, op.cit., p. 200.
10. J. A. Schumpeter, *Capitalism, Socialism and Democracy* (London: Allen & Unwin, 1954), pp. 84–5.
11. J. A. Schumpeter, 'The Analysis of Economic Change', in the *Review of Economics and Statistics*, May 1935, p. 7.
12. C. Freeman, 'Introduction', in Dosi, op.cit., p. 5.
13. See, for example, Perez and Soete in Dosi, op.cit., pp. 458–79.
14. This section draws heavily on Ch. 3 of James Pickett, *Agriculture, the Market and the State: Economic Development in Ethiopia*, OECD, Paris, 1990.
15. For a much fuller discussion of the characteristics and performance of the African economies, see, James Pickett, 'The Low-Income Economies of Sub-Saharan Africa: Problems and Prospects', in Pickett and Singer (eds.), *Towards Economic Recovery in Africa* (London: Routledge, 1990), pp. 215–66.
16. Henry Sidgwick, *The Principles of Political Economy* (London: Macmillan, 1887).
17. For a discussion of total factor productivity growth see E. Shaeeldin, 'Total Factor Productivity Growth in Developing Africa', in *African Development Review*, July 1989.
18. Adam Smith, *Wealth of Nations,* published 1776 (New York, Modern Library, 1937), p. 359.
19. W. A. Lewis, *Industrialization in the Gold Coast*, Government Printer, Accra, 1953.
20. See, for example, John Mellor, *The New Economics of Growth* (New York: Cornell, 1975).
21. R. Findlay, 'Trade, Development and the State', in Ranis and Schultz (eds.), *The State of Development Economics* (Oxford: Blackwell, 1988), pp. 78–99.

Technology from the People: Technology Transfer and Indigenous Knowledge

HELEN APPLETON AND ANDY JEANS

Discussions of technology transfer tend to be dominated by top down considerations of which hardware to choose and of how to best deliver the goods from developed to developing countries. Northern governments tend to link 'aid' with 'trade' and regard supply of hardware as an integral part of the aid package. But this approach is deficient in three respects: it ignores existing local capacity and expertise; it neglects the importance of working with and strengthening that capacity, and it fails to incorporate local priorities and decision-making frameworks. On a daily basis in countries of the South, the most useful goods often come not by air or sea but on foot from the nearest market place, and not as technical assistance but as part of a process of local innovation which uses local materials and incorporates local constraints. As Schumacher said, one needs to 'find out what people are doing, and help them to do it better'. This paper[1] examines some of the links between technology transfer and indigenous knowledge, and argues that the former is only valid when it builds on the latter and aims to reduce dependency and support self-sustainability.

In 1982 the Chinese National Offshore Oil Corporation signed a series of joint venture agreements with foreign oil companies. By 1984 relations between CNOOC and the oil companies had turned sour. The oil companies were frustrated, because they felt they were acting in good faith in training the Chinese to run the offshore oil facilities, teaching local workers to carry out the tasks performed by expatriates. This was what 'technology transfer' meant to the foreign companies.

Helen Appleton and Andy Jeans are with the Intermediate Technology Development Group, Rugby, England.

Unfortunately, this was not what the Chinese meant when they asked for the transfer of technology. The Chinese called the knowledge of how to operate installed systems 'know-how', which was not the essence of technology transfer. This essence resided instead in 'know why', the understanding of the scientific and technological principles underlying the design of the machines. 'Know why' contained an appreciation of why the system was designed the way it was: why specific machines were used, and why others were not, why operations tasks were the way they were, and why operator training was structured in the way it was. Without 'know why', the Chinese felt they could not take up the new technology, and complete the transfer process.[2]

This example illustrates the problems developed country organisations can create in their relations with developing country partners as a result of an incomplete conception of what technology transfer is about. Just as 'technology' is much more than a product or a process, 'technology transfer' involves far more than just moving a product or a process to a new place – and more than creating a factory staffed with local operators in a new place. Technology transfer should mean the exchange of the capability, and the thinking behind the capability, both to enhance existing capacity and to support Third World organisations in their own design and development efforts.

What is Technology Transfer?

Technology transfer is a process of sharing products, skills and ideas that takes place over a period of time. It is not a one-way event, with expertise passing from Europe to developing countries. The most successful transfers are those which see the greatest exchange between the two parties. Developed country agencies may have a better awareness of technical possibilities owing to better access to information flows within international scientific and technological communications networks. However, developing country organisations usually have a better grasp of local resource and market conditions, local priorities and local capacity to maintain production throughout fluctuations in local economies. Similarly they are more aware of, and better able to maintain, communication with other producers which is necessary for sharing product and process development on a wider basis, and essential for building a self sustainable technological base.

For example, several small British firms have been working with Nepali manufacturers on the design of low-cost micro-hydroelectric systems for rural mills. The UK firms' knowledge of new turbine designs, and in particular of the potential use of electronic control systems in governing, has provided lower cost components for their partners. But the Nepalis are by no means passive recipients of these products and ideas. Aware of the financial limitations of most mill owners, local manufacturers have identified other possibilities for low-wattage system (less than 10 kW) governing, using locally available induction motors. This new technical innovation (still being researched) could greatly expand the available market for electricity production equipment, providing new work for all parties. Benefits for Nepal would derive from a thorough exchange of knowledge and support for local capacity to both design and assemble systems which better reflect local constraints, and also to identify new opportunities for technical development.

What Kinds of Technology can be Transferred?

The simple answer to this is that all kinds of technology can be transferred, but for *successful* transfer there has to be an understanding of what local people want and what local people are already doing. There are many cases in which the most appropriate technology is one that is 'simpler', in terms of requiring less technical skill and understanding. For example, it is pointless to try to run village canning operations that require a trained food technologist to ensure that sanitary conditions are maintained, if there are no food technologists in the village. On the other hand, most acid preservation processes (making jams and chutneys, for example), can be run and quality tested by villagers given very basic training and simple refractometer testing equipment.

On other occasions the most appropriate technology for developing country production is a more 'high-tech' option. In the electrification of small mills in Nepal, sophisticated electronics technology is proving far more useful and transferable than simpler mechanical governing devices. This is because while the mechanical devices are simpler in terms of being older and more familiar, they are far more complicated to operate and maintain, needing trained engineer operators to keep them performing properly. Engineers are not easy to come by, however, in villages several days walk from the nearest road. Electronic load controllers, by contrast, are operated successfully by the mill owners themselves. They require no maintenance. Repairs require only a go/no go decision on replacing circuit boards.[3]

Developed country organisations should not automatically assume, therefore, that 'appropriate' products and equipment for developing countries have to be either 'simple' or 'obsolete' in Western terms. In general, developing countries possess larger resources of low-cost labour than developed countries, so technologies that are more labour-intensive are often the best choice for developing country manufacturing ventures. Similarly, because of the small, segmented nature of domestic markets in most developing nations, smaller-scale production units usually are more economical.

But it is dangerous to make generalisations on either of these counts. Both large- and small-scale sugar producers in western Kenya are finding that, while there are many unemployed in their area, there are few applicants for labouring jobs on the production lines. This is because many of the unemployed have received some education and training, and do not wish to perform low-paid, low-skill tasks. National labour statistics may not indicate local availability for specific tasks, and in some cases labour-saving devices may be the best option.

Similarly, certain types of operation may not make sense on a scaled down basis, despite the presence of dispersed markets. Bottling enterprises would not be easier or more viable if they were smaller, or decentralised. Village electrification through decentralised power production makes sense only when extension of the larger, central grid system is prohibitively expensive. In many post-harvest processing operations such as oil expelling and maize milling, the most viable course is to seek an intermediate level of activity between village women's manual efforts and high-capacity developed country style plants.

Therefore, developed country organisations interested in providing technical assistance to developing countries must first critically examine their own assumptions behind the choice of technology incorporated into the package. Assumptions that low tech, labour-intensive operations are the most appropriate may be correct for some countries and production sectors, but wrong for others. Similarly, a more sophisticated technology may also be inappropriate, not because it is sophisticated *per se*, but it has been developed in response to differently perceived socio-economic needs.

The ITDG Perspective

ITDG, as a technology development agency, is committed to generating benefits for poor, usually rural, small-scale technology users and producers, through technical assistance and technology transfer. Work at this level tends to focus on very small-scale, often informal sector enterprises, and is substantially different to work with larger enterprises at the more visible end of the formal sector. The specific implications of these differences will be discussed later but it should be emphasised here that different scales of production require very different modes of operation and that small producers in developing countries operate under very different conditions and constraints.

A New Approach – Technology Transfer from Within

Technical transfer from North to South may not always be what developing countries require. What are often ignored by expatriate technical assistance specialists are the vast resources of knowledge, skills and innovation already existing within small enterprises in developing countries, and the fact that appropriate assistance in many cases involves improving the distribution of information about existing products, processes, techniques and materials. There is much to be gained from transferring local expertise to wider audiences of small producers within countries and regions, as indigenous innovations, or 'peoples' technologies', can play a major role in national and regional economies. They are responsible for providing cassava food products to the majority of urban households in West Africa, and are maintaining the livelihood of many artisanal fishermen in the Bay of Bengal and Arabian Sea coastal communities. Where the informal sector is thriving, it is doing so not because of scientists, engineers, foreign technical experts or technical assistance programmes, but because of the activities of artisan innovators and the customers that advise them.

People's technology transfer has been limited by lack of awareness of its existence and a lack of understanding of constraints to its operation on the part of both governments and donor agencies. ITDG has supported investigations by researchers from 14 developing countries in Asia, Africa and Latin America into successful local innovation, which is providing a clearer picture of how these new technologies develop and spread, and of the factors that support and inhibit this process. Some of the case material is summarised below.[4]

Cassava Processing in Nigeria[5]

Gari, which is an edible product of cassava tubers, has become particularly important to Nigeria recently because of a ban on the import of cheap wheat, maize and rice. Cassava processing is now a major business, providing employment and income for farmers, processors and the producers of processing equipment.

Gari production has been helped by the development, by informal sector workers and R&D institutions, of grating, dewatering, sieving and frying machines. The most successful innovations in processing have come from the small-scale artisan manufacturers, who, by copying formal sector designs, have produced cheaper and more easily maintained equipment from locally available materials. Small producers have been able to respond to specific customer needs and to make modifications based on user experiences and suggestions. For example, production of sturdier machines and more durable grater sheets have been prompted by the demands of women users, who well understand the technical implications of *gari* production, although their relative lack of access to credit means that they may be losing control of the process. These innovators, however, like many others, are constrained by a lack of basic tools and infrastructure, and by a lack of access to credit and to information about products and materials.

Artisan Fishing Gear in Kerala State, India[6]

In India, artisanal fishermen have not only pioneered the design of small fishing craft and equipment still in use today, but also are continuously involved in developing new technologies and adapting them to their local marine environment. The fishermen of Kanyakumari and Trivandrum districts have developed highly skilful hook and line fishing and kattumaram sailing techniques to cope with the steep, uneven, and surf-ridden inshore waters of the Arabian Sea. They have continued to refine their skills (switching to synthetic fishing lines, for example, when these became available in the 1950s), and until the late 1970s their innovations contributed to a steadily increasing output from Kerala state's fisheries.

However, the introduction of large-scale bottom trawling in the 1960s damaged the shallow water marine environment and greatly depleted fish populations, and by the late 1970s the region's fisheries production, especially artisanal fishing, began to decline. In response, the artisanal fishermen explored ways to recreate reefs, experimenting with new materials and modifying reef designs to protect reefs from drift net entanglement. Appropriate resources directed at the fishermen's organisation – the South Indian Federation of Fishworkers Societies – enabled them to support the experiments and to improve distribution of information about innovations among local organisations.

These and other artisan innovations have preserved the livelihoods of fishermen and their communities. In addition, fisheries research institutions, which formerly had little interest in the artisan sector, now collaborate with fishermen's associations, sharing information and resources, which encourages the further improvement of reefs, bait and other essential small-scale fishing technologies.

People's Housing Construction in Huancayo, Peru[7]

Huancayo, Peru, is a city of about 200,000 inhabitants where recently areas of land have been expropriated for squatter housing settlements, which are lacking in all basic facilities. The majority of squatters are migrants from peasant communities and are undergoing rapid social change, a process which influences the knowledge that they have about the design of shelter and the technical aspects of construction. The fact that the shelter is temporary also influences peoples' choice of building materials. In rural areas building materials such as clay and thatching grass are free, but in towns many materials have to be purchased.

Traditional knowledge specifies the soils which can be used for adobe walls and the value of various strengthening additives. Circumstances have prompted modifications in the course of time, however, such as a reduction in the size of bricks, and the use of soil without strengtheners. Such modifications may not be technical 'improvements' in Western eyes, but neither do they indicate resistance to change or lack of technical knowledge. Instead they represent a logical response to the conditions of extreme poverty in which people live. For example, *tapial*, knowledge of which was originally confined to the high Andes, has proved to be a cheap and straightforward technology in situations of temporary housing at lower altitudes, and has become widely adopted without external assistance. Its use does not conflict with peoples' aspirations towards more permanent housing of stronger materials. This concept of 'temporary-permanent' housing underlies the fact that poor people too have several priorities in their lives, of which housing is only one.

How to Respond to Local Circumstances?

Industrial specialists and economists in Northern countries are well aware that different scales of production require very different modes of operation and encounter very different problems. Development agencies must appreciate that small producers in developing countries also operate under very different conditions, and that technical assistance has to be formulated around very different constraints.

For a start, few rural inhabitants are full-time producers. Most spend certain parts of the year fully occupied with subsistence (or cash crop) production, with off-farm manufacturing providing work and income opportunities during periods of low field labour needs. Even in urban areas where much production takes place in the informal sector, there is often movement between several activities within a household, rather than total dependence on one, and women particularly may be involved in several different activities over the course of a single day.

Because of this, developing country small producers can have very different economic objectives to their developed country counterparts. Maximising production and profit over a certain level could conflict with agricultural needs or other economic interests. Nor is production efficiency (maximising output per unit of time) always a major goal. Extra output can demand extra productive activity in other areas which may simply not be available given the time constraints of existing domestic, agricultural and other activities.

Another important difference between developed and developing country small producers is that many more of the latter are women. The smaller the scale of production, the more women tend to predominate, particularly in post-harvest processing enterprises. Aid efforts and corporate twinning arrangements often overlook this fact, with the result that the developing country producers with the most skills and experience are excluded from the technology transfer process. Even when women producers are considered, the constraints on their time imposed by the number of tasks that they daily perform, and the constraints arising from women's position in particular societies, are not recognised.

Informal sector producers in developing countries also have to contend with lack of access to basic facilities which developed country producers, and even larger formal sector workers in their own countries, take for granted. Many developing country artisans are illiterate and further educational and training opportunities are limited. Credit provision is usually poor: there are far fewer credit channels available, and high interest rates and short payback periods are required. This means that those who are illiterate and who have no collateral to offer (which applies particularly to women) will either find that credit is non-existent or will be forced to use informal moneylenders charging high interest rates. Foreign exchange can be difficult or impossible to obtain on a regular basis, and alternative sources of capital, such as venture finance, are not commonly found. Studies have shown that 95 per cent of developing country producers must rely on their own and family savings to capitalise their operations.[8]

Similarly the necessary infrastructure for small producer activity is also lacking: access to roads and markets may be poor or non-existent; electricity an unimagined luxury, and information about alternative materials, other products, new developments and markets, is unobtainable. In many cases the information exists but is poorly distributed, or is distributed in a form that renders it useless or inaccessible to small producers. The fact that small producers and entrepreneurs must have access to such information in order to upgrade products and processes is often not recognised.

Development agencies and Northern governments should also appreciate that developing country producers operate within different fiscal and policy contexts which may hinder the development of local technical knowledge. For many small-scale artisans the process of registering a business can involve endless and expensive formalities which may mean that enterprises are simply not registered. Through not being registered, such enterprises may not be eligible for grants and loans, or may not have access to training courses or other sources of advice.

Other aspects of government policy can hinder technology transfer and local technology development. The rapid growth of urban areas in many developing countries has brought a great need for low cost housing. This, in turn, provides a considerable opportunity for the production of low cost building materials, like pozzolana cements, fibre concrete tiles, rammed earth bricks, etc. However, the market for such materials often is constrained by building codes left over from colonial periods, which mandate the use of Portland cements even in low-rise buildings, require roof structures to be able to withstand six feet of snow, and pose other technically absurd barriers. The technology transfer process needs to include work on improving government understanding of what is necessary in the national context, and on modifying codes and regulations to stimulate the local production of technological alternatives.

A final factor for the developed country organisations to consider is time. No matter how great the technical and market opportunity, the process of technology transfer takes time. Each country will have different priorities and different levels of economic and technical knowledge and expertise which will require different adaptations of whichever technology it is proposed to transfer. For example, rainwater harvesting work among the Turkana nomads in northern Kenya was preceded by at least eight months of apparent technical 'inactivity', during which time project officers simply talked to people about locally perceived needs, discussed choices, and worked with locals over organisation, adaptation and control of the available techniques. Similarly, in Sri Lanka, the revival to micro-hydrotechnology took three years to pilot project activity before becoming accepted, and only after seven years is the local capability sufficiently consolidated to generate initiatives for moving the technology from the tea estates into the villages.

Conclusions

For development agencies and both Northern and Southern governments interested in technology transfer, the case studies and discussion illustrate three main points: first, that technical knowledge and a willingness to improvise and innovate exist within the informal sector and at grass roots level; second, that if technical assistance is to bring about sustainable development it should concentrate on encouraging and supporting this capacity; and third, that an important element of technology transfer is an understanding of the total policy environment into which it is intended that the technology be introduced, and a willingness to address issues that arise from that environment, such as lack of access to credit, training, information and infrastructure.

If development agencies are concerned about technical transfer helping poorer communities, they have to ensure that the benefits reach the people in these communities. In order to do this, poorer producers, men and women, should be included in the design and development stages of technology transfer. Improved information networks are needed both to facilitate the South–South distribution of knowledge and experience and better to focus the efforts of Northern agencies. Technical development programmes should reflect these objectives: work on the technology should be balanced by work among existing producers; thoughts of short-term technical success should be balanced by ideas of longer-term technical sustainability, and work at project level should be balanced by an understanding of the wider context. The valuable potential that exists is not being realised because governments and aid organisations have been unaware of the scale and the inventiveness of human activity, especially at the grassroots level. 'Top down' has to be superceded by 'bottom up' – 'find out what people are doing and help them to do it better'.

NOTES

1. This paper is partly based on a paper by Gamser, Holland and Appleton, 'Technology and Small Enterprise: Rethinking the Transfer Paradigm', prepared for the Conference on Small Enterprise Development in Oslo in June 1989.
2. Case example taken from Geoffrey Oldham, 'The Transfer of Technology to Developing Countries,' *Appropriate Technology* 14:3 (Dec. 1987).
3. Drummond Hislop, 'The Micro-Hydro Programme in Nepal – A Case Study', in Marilyn Carr (ed.), *Sustainable Industrial Development* (London: IT Publications, 1988).
4. The book containing all the case studies and discussion is M. Gamser, H. Appleton, N. Carter, *Tinker Tiller Technical Change* (London: IT Publications, 1990).
5. For more information, see Selina Adjebeng-Asem, 'User Demand and Product Inno-

vation', and R. O. Adegboye and J. A. Akinwumi, 'Intermediate Cassava Processing Technology Innovations in Parts of Oyo and Bendel States of Nigeria', in Gamser, Appleton and Carter.
6. For more information, see John Kadappuram, 'Peoples' Technology in Fisheries, India', in Gamser, Appleton and Carter.
7. For more information, see F. Monzon, 'Peoples' Innovations in Housing Construction in Huancayo, Peru', in Gamser, Appleton and Carter.
8. Carl Liedholm and Donald Mead, *Small Scale Industries in Developing Countries* (Lansing, MI: Michigan State University Press, 1987).

Technology Transfer or Technology Development? Third World Engineer's Dilemma

ENOCK MASANJA

The economies of many Third World countries are very weak; their manufacturing bases even weaker. A decade of low foreign exchange earnings has led to low levels of re-investment and to a general decline of manufacturing industries, with subsequent failure of the whole import substitution sector. Industrial growth has at best been small and slow; and not infrequently has given way to industrial decline. On the other hand, the export world market is very competitive and in practical terms, more restrictive to Third World products. Indeed, very few Third World industries operate at sufficient capacity and productivity to be able to compete on the world market.

The appropriate application of new advances in technological and material science can help improve Third World development, but choice of technology must take into account the competence and capability of the local pool of trained technical manpower and maintenance support. New technology, with its associated complexity and cost, must be deployed within the constraints of the indigenous technology capability, in particular those of manpower, technical support and maintenance capability. The technology must be able not only to sustain but also to improve both levels of production and the industrial capacity utilisation in existing and new plants. This paper points out some prerequisites for such technological development based on the author's working experience both in his home country, Tanzania and in the UK.

Introduction

North–South co-operation, has a diversity of perspectives which depend very much on where the individual, enterprise or government stands. In mentioning some historical facts below, I recognise that my perspectives

Enock Masanja is in the Chemical and Process Engineering Department, University of Dar-Es-Salaam, Tanzania. (Contact address: Chemical Engineering Department, Edinburgh University, Edinburgh EH9 3JL, Scotland.)

are not universally held. During the period of colonisation in the Third World countries, the large-scale exploitation of natural resources and labour was not in general accompanied by a similar scale of investment in any physical development sector, nor in the development of human capital in knowledge, health, social organisation, etc. Any development demands that all levels of manpower be adequately trained. This fact was completely neglected; so that the only type and level of education provided was based on providing manpower for clerical and petty managerial jobs: technical education was completely non-existent.

In my country, for example, as recently as 1982, that is, 20 years after independence, there were only about six graduate chemical engineers. There is still today a lack of technically qualified personnel who are competent to improve indigenous technologies, to acquire or develop new technologies, to develop and improve research both in quality and volume, to promote greater interaction between the users of technology (that is, industries) and the producers of technologies, and to promote a greater commitment to the development of technology at the level of governments and other legislating bodies.

A Third World Engineer's Difficulties and Dilemma

The difficulties faced in the practice of engineering in the Third World are enormous. I understand that similar difficulties face other technical professions as well. I have tried to list some of these problems below.

Purpose of Industry

The purpose of industry is to provide essential goods and to create wealth. For industry to prosper, the right environment must exist; this includes a properly trained and motivated manpower pool; and appropriate technology that is relevant to that environment. These factors must be understood and appreciated in their own local complexity.

In Third World countries, industry and technology must cater for two distinct constituencies: the nation, and the rural or village level. At rural or village level, industry and technology are required particularly to create employment, to provide services and to produce inputs to the agricultural sector. Equally importantly, it must help to alleviate the heavy workload done by women. Here, the priority of the required products might vary from one place to another, but is likely to include corn grinding units, water pumps, building materials and cooking stoves. By virtue of their locality and end use, the technology must be appropriate, required equipment or implements must be robust, easy to

use and maintain with minimum training and preferably depend on renewable energy like solar or wind power.

At national level, technology and industry must additionally address
the provision of goods for both the local and export markets. It would
be naive to even contemplate that these countries would be able to
earn substantially more foreign exchange by exporting traditional un-
processed agricultural products. Further the market is diminishing in size
as the consumer populations in the North continue to fall or remain fairly
constant, whereas those in the South continue to explode at an alarming
rate. In addition, producer prices on the world market have kept falling,
and efforts to increase agricultural production would in effect push the
prices even lower. And the South–South trade market has yet to be fully
exploited.

For a largely agricultural economy, industry must be able to transform
agriculture and expand its productive capacity in diversity of products
and in value, to improve productivity by using suitable farm implements;
and to improve harvest recovery and reduce post-harvest losses. Industry
must also support agricultural research aimed at developing new or
improved agricultural inputs such as irrigation systems, improved seeds,
fertilisers and pesticides.

There is probably a lack of commitment to practical self-reliance.
Also, poor collaboration amongst different professions, scarcity of the
necessary support staff such as technicians and artisans, and general lack
of qualified personnel, make the effort to improve industrial productivity
an uphill struggle. The absence of even basic tools and equipment such as
balances, thermometers, pH meters etc. contributes significantly to loss
of productivity.

Education and Training

For Third World industries to be able to keep abreast of the revolu-
tionary advances being made in information technology, biotechnology
and material science, and if the gap between the North and South is ever
to become narrower, the South must have a qualified workforce with
high-level training in all technical and professional fields. Technical education thus needs to be improved as an urgent priority. Technology
transfer to areas where there are no qualified technical personnel is
impractical and effectively impossible.

Although most Third World countries have revised their education
policies, poor economic performances have led to drastic cuts in funding
for education and training, for equipment, books, periodicals, research
and field work, and for physical facilities such as schools and laboratories.

There are pressing shortages too of technical books and periodicals, not only in schools and colleges but also in industries. Technical literature collections in national libraries are still very poor and will remain so for the foreseeable future. I was moved by therecent public book appeal for Romania and wished that there were more such charities for other Third World countries. The importance of such literature cannot be over emphasised.

Another inherent drawback of poor technical education in most Third World countries is often the lack of proper professional training, as the number of senior qualified and experienced personnel in industries is low and sometimes completely non-existent. It is not uncommon in my country for a fresh graduate to assume senior responsibilities on his first day of employment, and thus be required to assume responsibilities for which he or she is not properly trained. This is known to have had an adverse effect on graduates' morale and productivity. These graduates take a much longer time to develop their professional skills compared with their equivalents in the North who undergo rigorous professional training prior to assuming responsible jobs. This drawback hinders the exploitation of the graduates' understanding of the basic laws of physical science, and of their creativity and innovativeness.

Information and its Dissemination

Infrastructure is an important conduit for both information and goods. In most Third World countries, communication and general transport is very poor. This does not render any help to the already dismal co-operation between industries. Although most of the industries are owned by the government, still there is no sound coordination nor co-operation. It is not uncommon for one industry not to be aware of what the next industry's capabilities are. A local search for services or materials becomes difficult, and consequently there is a perpetual dependence on imports, even for goods and services that could be provided locally. It is probably on this issue that the absence of private sector initiative and free market forces is most evidently felt. There are few information exchange fora; organised seminars or workshops are few and irregular; and publications on industrial and research matters are more or less non-existent.

Appropriate Technology

In the long term, high technology, though initially expensive, has the capability to offer huge benefits in improved safety, reliability, energy efficiency, operability, flexibility, convenience, profit and productivity. Some of these benefits might not be evident initially, and all are highly dependent on accurate appraisal and choice. In order to be able to compete in the world market, product quality must be high and consistent, while production costs must always be pushed down. Third World countries must build that capability if they are ever to increase, or even maintain, their current level of foreign exchange earnings. As an engineer, I would advocate the use of high technology. For weak economies, however, such as those in the Third World, the choice demands considerations of the qualification and size of the technical manpower pool; maintenance requirements; and climatic factors.

Often, however, the choice is between the new and the old technology, or between 'high' and 'low' technology. The appropriate choice usually lies in neither extreme. This dilemma can be an expensive one and needs to be resolved by a thorough, rational and exhaustive appraisal and assessment of all options. Often there is a trade-off between technologies. Any one technology selected becomes appropriate if, and only, if it can be adapted to fit in the local environment. As no technology can ever be transferred by 'surgical knife techniques', local experts must be involved, and the success or failure of the technology transfer or development depends heavily on them.

Conclusion

It is probably difficult for someone from the Third World to address the issue of technology transfer without any sentiments. We experience many difficulties, and to some these are indeed difficult to understand let alone appreciate. In the last decade, the Third World's economies were in their infancy and, just like an infant learning to walk, occasional stumbles and scars were to be expected. There are a number of third world countries which have established reasonable foundations for technology acquisition or transfer.

It is to be hoped that their choice of technology will be dictated not only by the strength of their economy but by a strategy for developing a modified indigenous technological capability. The future, to me does not look to be that bleak! Third World industries desperately need some assistance. The basic area is likely to include establishing an information directory, organised at national and/or regional level, detailing information about local industries. The directory could include other information such as a data bank of chemical properties, safety, standards, sources of raw materials locally and regionally, lists of up-to-date industries and their activities, and some market information. The task of running such an information network could be entrusted to a research organisation or higher learning institution. Another assistance could be in developing an efficient network of professional associations in all technical and professional fields, organised both at national and regional level. This would form a forum for exchange of information on research and development through seminar and workshops or publication of periodicals (things that are not currently done at regular intervals). Through establishing professional, technician and artisan training; and through general support of physical science education – for example, to enhance science institutions and help to establish research facilities and capabilities. Lastly, there should be an international appreciation and understanding for the presence of a growing number of local experts who need to be consulted on technology transfer and related issues.

Finally, recent developments in South Africa give hope for an end to the costly destabilisation we have had to live with during the last 20 years or so. And I sincerely hope that current events in Eastern Europe will not mean a reduction (in real terms) of technical, economical and social aid to countries of the Third World.

Science and Technology in China: Transition and Dilemma

YAOXING HU

During the 1980s China's science and technology underwent substantial changes. These changes have seen a transition towards a rationalisation of the country's scientific and technical resources, but have also encountered many difficulties and pose a serious dilemma. This paper sets out to review these changes in the context of the Reform and Open-door Policy. These are illustrated from aspects concerning policy orientations in which national strategic programmes are devised, the dynamics of the research and production sectors, and the acquisition of foreign technology. The author is inclined to the view that the potential of technological capabilities and economic development is promising. But progress depends on further steps to challenge the conventional social and economic system.

Science and Technology Policy in the 1980s

The role of science and technology (S&T) in the economic development of China has always been an important issue among the top decision-makers. In terms of long-term strategy, however, the policies were not coherently stable. The situation was distinguished by conflicts between ambitious development goals and real economic constraints, and between biased investment in capital goods sectors and the popularisation of grass-root innovations.

Based on previous lessons, the leadership and the scientific community grapple with issues concerning priorities of various research and development (R&D) activities, and the best possible mechanisms for promoting industrial innovation and widespread assimilation of up-to-date technology. New approaches regarding S&T and economic development as a whole can be synthesised into several major shifts (or 'principles'); that is,

Yaoxing Hu is at the Science Policy Research Unit, University of Sussex, England.

TABLE 1
NATIONAL R&D PROGRAMMES (1986 ONWARDS)

Programmes	Commencing Date	Spending (million yuan)	S&T Personnel involved
Key Projects Plan	1986	3300 (1986–90)	100,000
Spark Programme	1986	3760 (1989)	
Dissemination Plan	1989	300 (1989)	
High-Tech Plan	1987	4700 (1987–89)	12,000
Torch Programme	1988	1500 (1988–89)	
		1800 (1989)	
Basic Research Plan	1986	124 (1989)	

from the S&T sector to the production sector; from the military to the civilian sectors; from the advantages of coastal cities to the interior lands; and from S&T abroad to homogeneous diffusion. This was to characterise the mainstream of development strategy in the 1980s, but reality is yet to evolve and in many respects it fell far short of intentions.

To comply with those principles, scientific research and related technical undertakings are defined as being threefold. First, the most important task is for R&D to attain the national goal of quadrupling GNP at the 1980 level by the end of the century – the priority areas being agriculture, energy, resources exploitation and utilisation, and studies related to population problems. The second task is concerned with following up the latest scientific and technological advances in the world. Third, basic research is to be continuously promoted. Since the mid-1980s various government initiatives have taken shape to reflect the above strategy. Six large-scale R&D programmes at the national level are now under way (Table 1).

R&D projects devoted to maintaining economic development are of primary importance and include the first three programmes. The Key Projects Plan was an instrument of the seventh Five-Year Plan (FYP, 1986–90), covering the technologies of major priority areas of the economy. The Spark Programme is intended for rural development through applications of recent research and related technical achievement. The S&T Dissemination Plan is a similar initiative for the application of new and high technologies to industrial sectors.

Some of the Torch projects are also to be integrated into these approaches. These initiatives will eventually enhance production capabilities; some of them have already represented an improved input/output ratio at 6.51:1.[1]

The High-Technology Plan (also called the '863' Plan) is aimed at following up the world trend of high technologies, consisting of training

TABLE 2

R&D ESTABLISHMENTS IN CHINA

Categories of R&D Units	Number	Sci./Eng.	Budget (m. yuan)
R&D Units at County Level	3360	6172	214.6
R&D Units Above County Level	4706	122432	4085.9
Ministerial Departments	932	175939	6247.8
Chinese Academy of Sciences	123	49865	866.3
R&D Units in Universities	429		

high quality personnel, preparing the economic sectors for the next century, and paving the way for selected high technology industries. A complementary initiative, the Torch Programme, is aimed at commercialising the R&D results in new technologies. Plans concerning the establishment and growth of relevant industries are under consideration.[2] Useful approaches have been taken, such as setting up new technology development zones in selected areas where research establishments and universities are concentrated.

Basic research is largely the responsibility of the Chinese Academy of Sciences (CAS) and the National Natural Science Foundation (NNSF). The latter evolved as an independent entity in 1986, with an annual budget of around 100 million yuan. Most of the NNSF activities are now incorporated into the Basic Research Plan. According to Song Jian, State Counsellor and Minister of State Science and Technology Commission (14 February 1989), basic research funding will increase in proportion to overall R&D appropriation. This means it will rise, by the end of the seventh FYP, from the present 7.1 per cent (in 1987 the actual amount of basic research expenditure was 800 million yuan) to 8–9 per cent of total R&D spending, and will further rise to around 10 per cent for the 1990s.

Emerging Role of Research Establishments

There are over 8 million S&T personnel in China, with nearly 355,000 full-time equivalent researchers in over 9,500 institutions (Table 2).[3] The research sector has been undergoing serious revamping in recent years. In order to illustrate various changes in these research institutions, let us look at the country's most comprehensive research organisation – the Chinese Academy of Sciences. The Academy's strength lies in basic and interdisciplinary research, and it serves as a trail blazer in developing those research areas where China was previously blank or weak. The institutes of the CAS are ranked large and medium-sized in the country,

TABLE 3
TECHNOLOGY TRANSFER PROFILE OF CAS, 1984–88

Year	Projects Completed	Applicable Items (A)	Transferred Items (B)	(B)/(A)
1984	1281	804	415	56.62
1985	1606	981	394	40.16
1986	1411	728	567	72.40
1987	1259	726	507	69.83
1988	1028	579	382	66.32

Source: Guangming Daily, 20 September 1989.

with a total staff of 82,326. In addition to the research units, the CAS has some 30 other institutions including nine factories, six libraries and a University of Science and Technology of China.

In line with the 'shifting' guide-lines, the restructured goals of the CAS have been to address the broad issue of how research centres of this sort – that is, largely basic research oriented – should play a role in the new environment. The changes in recent years have been impressive. According to the President of the CAS, Zhou Guangzhao, almost two-thirds of the Academy's force has been engaged upon making a direct economic contribution.[4] Their main activities can be categorised as follows.

First, many institutes have done more work for the national R&D programmes relevant to economic tasks. Some of them have participated in the importation of technology and popularisation. For instance, among the 1,411 research projects completed in 1986, over two-thirds of the projects were commissioned under economic planning and the rest were contracted by ministries and industrial firms. A more recent undertaking is the work for agricultural development in the Huang-Huai-Hai Plains, which include regions of five provinces in East China. Thirty institutes with 600 researchers have been working to transform the low-yielding land for higher output.[5]

Second, the CAS has set up various operations for technology-related business and co-operations. Their activities include transferring technology to industrial sectors, co-sponsoring economic projects, and establishing business operations requiring a solid scientific base. By 1987 CAS had set up long-term business partnerships with several thousand industrial enterprises. The technical contracts concluded with different regions and firms accounted for 76.15 million yuan by 1987. Over half of the projects completed by the CAS are either transferred to industries or taken over by local authorities (Table 3).

The business operations which have been set up involve both different

TABLE 4

SHARE OF FIRMS' FIXED CAPITAL AND OUTPUT IN CHINA (1989)

	Number	Fixed Capital	Output
State-Owned Industries	100,000		
Large-Scale	500	65 Per cent	40 Per cent
Medium-sized	9,400		
Collective & Private	7.4 million		

units within CAS and those external to the organisation. Since 1983 CAS has formed 400 research-intensive entities, including joint ventures with industrial enterprises. By 1988 the total sales revenue had risen to 1 billion yuan. In 1986 CAS exported products to 26 countries and regions. Earnings increased by 160 per cent over those of 1985. The Dongfang Corporation alone earned US$15.6 million between 1984 and 1986 from technology export activities.

Third, institutions concerned with natural resources, energy and the environment have been reorganised in large teams for an integrated survey of resources. They have accumulated an extensive amount of information for economic planning and resources exploitation. An environmental research centre has been set up in order to upgrade facilities in resources development and environmental protection. Scientists and engineers take an active part as well in basic research and high technology projects. Over 20 research labs and institutes have been opened up to the outside world and are attracting promising scientists. At the same time, scientific personnel are encouraged to explore research topics themselves to solve more realistic problems.

A long-term goal of the CAS proposed by its President is that it should become a centre of excellence composed of state laboratories (or institutes), state science engineering centres, resource and environment R&D centres, jointly managed open laboratories, and R&D institutions attached to high technology conglomerates.

Innovation and the Production Sector

China has already achieved a relatively comprehensive structure of industries, including mining, manufacturing and utilities. State-owned enterprises play a dominating role, accounting for 70.4 per cent (in 1985) of industrial output value. The large and medium-sized firms contributed 80 per cent of state revenue in 1989 (Table 4).[6]

The industrial sector has so far advanced along the road of increased investment in new plants. This expansion has resulted in structural imbalance and low productivity. Therefore the existing production capabilities continue to be at the expense of over-use of energy resources. This development model and the evolving structure has hindered inno-vations in various ways concerning investment decisions, operation of the factory, product quality, technology import, the assimilation of new technology, and the overall management style as well.

This situation is more obvious in the manufacturing industry, which requires a higher degree of sophistication; and the integration process of new technologies into production facilities is rather slow. There are about four million machine tools at present; high-quality and technically-sophisticated ones account for less than 10 per cent. Therefore, over 60 per cent of all manufactured goods are at the level of the 1960s.[7] Meanwhile, the degree to which research results can be applied to production is much less. According to a recent survey, no more than 15 per cent of research results have been utilised, and less than five per cent of patents filed have been applied to production.[8]

The importance of technological change and the need to up-grade industrial capabilities were recognised in the 1980s.[9] One of the means of bringing this about has been to make production firms into economic entities, more independent of government bodies, with greater autonomy in decision-making. But the practice has been somewhat limited so far. Industrial reforms have not released firms from their shortage of funds for innovation activities. When tax reforms were introduced in the mid-1980s firms were still burdened with various duties which fall outside the formal tax regime, so that little funds were left for technical up-grading.[10] According to the State Statistical Bureau (1989), the funds for technical transformation and equipment up-grading were only 78 billion yuan in 1989, which was 20 per cent less than in 1988. This is largely due to rigid quotas contracted to firms. Meanwhile, the improperly-transformed price and tax mechanisms leave managers with less choice in production plans.

Other aspects of the reform package involve some institutional changes. Experiments were carried out to try a new way of managing capital assets. Stock-based companies were set up by either transform-ing the existing enterprises or creating new ones. But, confined by the conventional economic system, the practice has been accompanied by serious distortions. Public assets can be improperly appraised and easily undervalued. Shares from profit-making firms do not carry risks. Besides, share-holders are also entitled to abandon shares to the issuing firm.[11] An alternative practice for institutional change is displayed in the industrial conglomerates which have emerged in recent years. Firms of similar interests get together and are bound by loosely defined contracts. Some large conglomerates consist of as many as over 50 organisations. This has, in some way, facilitated the link between research organisations and industry. Research organisations are encouraged to join these industrial co-operations. Some have become subsidiaries in large industrial groups, and some have become technical service centres in their relevant sector. As this development is very recent, its impact is still unknown.

Recent years have already seen specialised R&D activities on the increase in large and medium-sized firms.[12] The number of R&D units was 5,525 in 1987, and grew to 8,004 in 1989. But only 20 per cent of all technical personnel in these firms actually work in R&D, whereas in developed economies this share is between 50 to 60 per cent. The scarcity of R&D personnel in industry is not only caused by the industrial structure but is also related to the educational policy and the country's personnel management. University graduates tend to go to government offices rather than production sites. Many students now studying abroad come directly from universities and other academic institutions. According to the present arrangement they all return to teaching and research posts when they have finished their studies. China needs more expertise in industrial techniques as well as managerial skills, rather than large numbers of highly trained professionals to pursue fashionable research topics. There is obviously a need for a rational mechanism to link these students with industrial organisations.

Technology Acquisition from Abroad

In order to modernise the country, over the last 40 years some US$30 billion has been spent on importing technology, including complete sets of equipment and turn-key production plants.[13] Up to the mid-1980s the technology acquired from abroad still largely consisted of complete sets of equipment and plants.[14] Among all the contracts signed for importing technology between 1973 and 1983, it accounted for over 95 per cent.[15]

Past experience shows that a policy of importing technology did not produce the anticipated results. There were inadequate procedures to verify the standard of technology, as well as commodity specifications, in the contracts. There were cases of disappointment and frustration, stemming directly from ambiguous contractual terms and owing to lack of understanding of international practice in relevant industries – for example, in technology transfer between the China National Off-shore Oil Corporation and the foreign oil companies.[16] Other problems were caused by lack of planning and co-ordination, and this has resulted in massive duplication in importing large numbers of similar production lines – for example, for colour televisions and refrigerators.[17] Some machinery was made redundant even before it was finally installed because the output capacity far exceeded the potential consumer demand. Meanwhile the cost of servicing these foreign technologies is high. One Jiangsu–Swedish joint venture established with US$12 million of investment has been totally reliant on Swedish equipment and raw materials for the last decade, hindering China from developing her products with domestic materials.[18]

Difficulties can also be caused by neglect of the available expertise and inability to cope with the process of decentralisation in decision making. There was no proper mechanism to control individuals who were given increasing autonomy in managing public assets. There were reported cases that dissatisfactory goods and equipment or even those 'Made in China' were imported. Such phenomena exist in the whole spectrum of foreign trade. Around 14.51 per cent of commodities imported in 1988 from major trading partners were found to be sub-standard. Claims against exporters mounted to $US100 million last year.[19] The ultimate reason for this is rooted in the dilemma between state ownership and decentralised management – a continuing contradiction in terms of the reform itself.

Since the mid-1980s there has been a growing consensus that the existing practice should not continue, while local conditions should be considered and home-based infant industries need to be protected. The emphasis therefore began to shift in favour of the 'software' aspect of technology, focusing on technical know-how and so on. But how far this objective will be achieved remains a question as long as the ownership system remains the same.

Conclusion

I have briefly discussed China's science and technology system, which is still under transformation. Promising changes are to be found in the country's effort to further strengthen the link between the S&T system and the production sectors. Some specific measures in the experiment stand out: these include the contract practice in order to facilitate technology transfer between institutions, the integration between research and industrial sectors, the public bidding for projects, and the priorities given to new and high technologies. Future development depends on where the wind blows in restructuring the economy – whether it is still centrally-controlled or market-oriented. The present difficulties and obstacles are rooted in the whole economic system itself, within which there is no mechanism for rational planning and adjustment. This is a serious challenge to the present framework of reform.

NOTES

1. 'National S&T Popularisation Programme under Implementation', *Guangming Daily*, 25 Feb. 1990.
2. Institute of S&T Information of China, *Popularise high technology in Chinese industries* (Beijing: The State Council, May 1989).
3. The figures are based on 1986, and do not include those in the social science fields. See State Science & Technology Commission, *Guide-lines of Science and Technology Policy in China* (Beijing: Science and Technology Documentation Press, 1987), pp.196–9, 239–41.
4. *People's Daily* (Overseas Edition), 30 March 1987.
5. Chinese Academy of Sciences(CAS), 'Agricultural Programme Management Bulletin', 24 Oct. 1988 (Internal Report).
6. *Economic Daily*, 13 and 19 Oct. 1989.
7. Lü Dong, 'Historical shift confronted by the Chinese Industry', *Liao Wang Weekly*, No. 2 (8 Jan. 1990), pp.9–10; Another source revealed that the share of numerical-controlled machine tools among the rest in Japan is 30 per cent, and 37 per cent in the USSR; but 0.03 per cent in China. See *Guangming Daily*, 13 Jan. 1989.
8. *Guangming Daily*, 29 March 1989.
9. Lü Dong, op. cit.
10. For detailed review on this issue, See Chen Tong and Jiang Xiaowei, 'Analysis on the Shortage of Funds in State-owned Industrial Enterprises', *Economic Research*, No.10 (Oct. 1987).
11. Zhang Yanning, Director of State Commission for Restructuring the Economic System, in *Economic Daily*, 21 Feb. 1989.
12. State Statistical Bureau, 'Statistical Communique on National Economy and Social Development, 1989', *People's Daily* (Overseas Edition), 22 Feb. 1990, p.3.
13. Chen Ping, 'Problems on Technology Import', *Economic Research*, No. 12 (1988), p.61.
14. Zhu Rongji, 'On the importation of technology: China's recent experience and policy', in R. Lalkaka and Wu Mingyu (eds.), *Managing Science Policy and Technology Acquisition: Strategies for China and a Changing World* (New York: UNFSSTD, 1983), pp.265–70.
15. Lin Guang, 'Issues on Increasing Economic Benefits of Technology Import', *Economic Theory and Business Management* (Beijing: People's University Press), No.5 (1983), pp.36–42.
16. Geoffrey Oldham *et al.*, *Technology Transfer to the Chinese Off-shore Industry* (Brighton: Science Policy Research Unit, 1987).
17. Gu Shulin, 'Discussion of Several Problems Concerning Technological Introductions from Abroad', in CAS Institute of Policy and Management, *Science and Technology in China – Selection from the Bulletin of the Chinese Academy of Sciences*, Vol. 2 (1987), pp.549–51.
18. *Beijing Review*, 13–26 Feb. 1989 p.39.
19. Zhu Zhenyuan, Director of the State Bureau for Inspection of Import and Export Commodities, ibid., p.9.

University Research in Britain for Rural Development in Africa: The Example of the Rural Ram Pump

TERRY THOMAS

Despite having some advantages such as skills, facilities and enthusiasm, Northern universities have difficulty in participating in the design and development of technologies for use in Africa because of difficulties in funding and in communicating with the potential users of such technologies. Warwick University has developed some ways of participating, as exemplified by a programme to reintroduce – on a basis of local manufacture – water-powered hydraulic ram pumps into rural Africa. New designs, materials, methods of manufacture and the organisation of installation have been explored in partnership with several indigenous organisations.

Introduction

In British universities there are many pockets of interest and goodwill concerning African development. This paper describes one way they are being harnessed, and discusses some of the conditions necessary for such a Northern contribution to Southern development to be useful.

When it comes to engaging in overseas development, universities have both strengths and weaknesses. They contain skilled people, and in the case of students 'cheap' semi-skilled ones. Their information base is good and in applied subjects they are well supplied with suitable equipment. University staff have some freedom to explore new topics or ones that can never be fully commercial; international links are strong; credibility is generally good (perhaps unjustifiably so). On the debit side, however, is the bias towards 'publishable research' which usually emphasises the general and abstract at the expense of the particular. In engineering for

Terry Thomas heads the Development Technology Unit, University of Warwick, England.

example, design is rather looked down on, being thought too localised and specific in its concerns. Universities ultimately live by their teaching, so that very rarely are senior staff free to concentrate wholly on scholarship, research or research management. Indeed at certain times of the year, teaching requires almost the undivided attention of the faculty members. There are no product-lines to pay for design and development work, so each substantial programme has to attract outside funding. Chasing grants and contracts seems to take up the bulk of the 'research time' of senior staff, who are often ill-organised for this task.

Development research poses particular difficulties, because of the great geographical and cultural distance between the researcher and the ultimate client group. Given the extra expenses of travel, and the high level of salaries in Britain compared with poorer countries, can UK involvement in development issues be cost-effective? The major sources of money for such studies are financially strained charities and governmental aid organisations rather than industry, research councils or generously funded ministries like defence.

Traditional 'development studies' have been mainly socio-economic. Professor Chambers [1983] quite entertainingly analyses the three-way relationships between social science commentators, rural development practitioners and the rural poor themselves. There are, however, also professional studies departments in British higher education specialising in the application to tropical countries of medicine, town-planning, transport, agriculture and so on. The staff of such departments globe-trot vigorously, especially in vacations and as consultants. Many of their students (usually postgraduate) come from Africa and Asia.

Warwick University's Development Technology Unit

Warwick University's large Engineering Department is by contrast mostly UK orientated. However, a decade ago it started a course for idealistic mechanical engineering students (B.Eng. Design and Appropriate Technology) in which professional studies are given a special focus on overseas development and on sustainable technologies for Britain. Some 130 students (out of over 1,000 taking engineering degrees in the same period) have entered the course since it started. One of the more positive outcomes has been the formation of a loose network of past and present students having interests in developing countries: around ten are currently in Africa. This has led in turn to the establishment of a Development Technology Unit to focus design activities of Warwick students, research students, graduates and staff on what has come to be called rural industrialisation.

So what contribution can an academic group like this make to African development? How can it use its skills and facilities? How can it compensate for its but partial knowledge of the distant potential users of its research efforts?

The Unit operates both in the 'demand-pull' and 'technology-push' modes. It receives enquiries by post and (more commonly) through questions and comment directed at its staff when in the field. It tries in consequence to identify problems of economic or social significance that it is competent to help resolve. However once it has engaged in a technical development programme, a degree of designer enthusiasm creeps in and some unsolicited demonstration of technology takes place. Writings on developmental topics often put forward the proposition that all technological change should be market or needs led. In practice there must be some dialogue between needs and technology, because needs are too often expressed only in terms of the means of satisfying them known to the needy community. 'Refining the specification' is a key part of any design process, and is properly influenced by a knowledge of possible solutions.

In practice it is difficult to refuse to respond to an enquiry, and any appropriate technology organisation can easily dissipate itself on low-level information provision. An innocent-looking question (could we in Kenya substitute concrete flywheels for cast-iron ones when copying Indian machines since casting is poorly developed in Kenya?) may need substantial research to back an authoritative answer. Intermediate technology, as a specialism, suffers seriously from amateurism and a failure to distinguish between the theoretical and practical performance of equipment: because funding is usually inadequate by normal industrial standards, new technologies are rarely properly designed or adequately tested.

The DTU would like to operate in a particular way. It would like to work through indigenous organisations in developing countries that are capable of identifying needs, measuring markets, developing specifications, testing new products or processes and performing the many components of the activity known as technology transfer. However, 'grass-roots' organisations are rarely that competent, and in Africa they are especially weak. Governmental organisations are dominant in sub-Saharan Africa, but have been becoming increasingly ineffective in recent years. Commercial firms are rarely in touch with, or even interested in, rural development. Farmers, small artisans and their organisations are difficult to communicate with from outside.

In practice, therefore, the DTU itself becomes involved in activities (like technology transfer) for which it is rather ill-suited. In one country of operation – Zambia – it has decided its best contribution would be to actually found a model rural technology centre. In others it places engineering students to act as local intermediaries in the needs/ possibilities dialogue.

Against this background I should like to describe a specific product-development programme that started five years ago and will take at least two more to complete. The example is that of the hydraulic ram water pump (which with rural radiotelephony and heat-powered ice-making constitute the major current research topics of the Unit).

The Hydram

Background

The hydram is a *water-powered* pump, invented about two hundred years ago and available from several manufacturers round the world [*Tools for Agriculture 1986*]. Conventionally it is made of cast steel with a steel drive pipe. Two designs for do-it-yourself hydrams have been published (ITDG, VITA) but both have very low capacity. Several studies of the pump exist in the hydraulic literature; the underlying mechanism of operation by multiple shock waves is fairly well understood.

It came therefore as a considerable surprise to have the hydram identified by rural development activists in five of six African countries visited in 1985 as a needed device. It transpired that it is a 'remembered' technology from former times, no longer available because it is too costly as an import and lacks experienced installers. Traditionally used for institutional water supplies, current interest also extends to its application in village water supplies and low-lift irrigation.

In consequence of this expressed interest, initial specifications were drawn up for two machines. For high lifts (say 50 metres) a steel machine is needed, compatible with a 50-mm diameter drive pipe. For low-lift irrigation a 100-mm plastic machine is more suitable. In both cases the drive pipe, not the pump, should constitute the main limit on throughput. For many reasons, in-country manufacture using readily available materials, is highly desirable.

Between 1985 and 1987 these pump specifications were refined and several early prototype machines were produced using 'student-power'. Commercial and published designs were examined and a very limited computer simulation was written. Crude test facilities were established in the University laboratories and in mid-Wales.

Design and Development

The Unit adopted the hydram as one of its major design projects, believing it to be suitable for university attention because:

- design innovation is required to meet the specification above,
- advanced techniques such as computer simulation appear likely to allow a simplification of design rules and the pretesting of major changes in geometry,
- instrumented laboratory testing should help calibrate simulations, show up design faults and demonstrate how to tune for high efficiencies and throughputs.

However, the university facilities are far from sufficient. Field testing of durability is required. The detailed requirements of African users and manufacturers have somehow to determine the development of new machines, and various approaches have been made to obtain this information.

A small farmer in Western Kenya was given the materials for a machine, but contact with him was then lost. A prototype irrigation hydram was built at an agricultural training centre in Zimbabwe; it later starred in an exhibition but has never been used for farming. A small high-lift steel machine was built and demonstrated at a rural hospital in Zaire: within a week news of it had spread a hundred kilometres. In each of these cases the installation agent was a student on vacation.

The feedback from these initial demonstrations indicated a need for much more thorough technical development in Britain, for closer linkages with African field organisations and for the later formation of new organisations to handle the manufacture, surveying, installation and maintenance functions that a successful hydram usage programme entails. It became clear that village water supply is institutionally much more complex than small-farm irrigation, but that even for the latter it is naive to treat a hydram pump simply as an item to be bought in a store.

Two further years brought modest advances on all fronts. Prototype machines were installed in a country park in England and a research station in Spain. An aid agency commissioned a demonstration of the irrigation variant in Zambia. An improved machine has replaced the first prototype in Zimbabwe. Three water supplies, for respectively a village, a school and a theological college, have been installed in Zaire using charitable funding. At Warwick simulation and laboratory test facilities have been expanded, and the understanding of both pump and system design improved. However this product-development work has been undertaken largely on short-term finance, by using volunteers or by taking (legitimate) advantage of existing university facilities.

In 1989 both the UK design work and the Zairian dissemination programme received longer term (that is, three-year) support, respectively from the Overseas Development Administration and the charity Tear Fund. In addition the University is attempting to found the rural technology centre (mentioned above) in Zambia and to intensify relations with an existing agricultural training centre in Zimbabwe. Hydrams will be one of several 'product packages' to be tested and propagated from these centres.

The hydram programme has thus developed considerable momentum and those working on it are busy solving the foreseen and unforeseen problems it has thrown up. Since the programme began, new constraints have arisen: for example the availability of steel and plastic piping (the raw materials for hydram manufacture) has dramatically worsened in several African countries in the last year. The better understanding that comes from any research reveals new contingencies to be designed for – theft, corrosion, poor manufacturing quality control, excessive expectations of users, inadequate organisation of payment and maintenance, seasonality of sales, flood damage, disputed ownership of water sources and so on.

There is a real danger of identifying too strongly with the particular technology. For that reason the Unit has recently made a specific effort to support partner organisations in using and assessing competing technologies – such as human-powered pumping or gravity-fed irrigation. Technology choice should precede technology transfer. There are, moreover, at least two groups usually involved at the receiving end of a technology transfer process, namely makers and users. In this case makers further divide into manufacturers and installers/builders, and users into farmers and households. For none of these will hydram pumps exist in isolation, but will be made or used alongside other devices. Even viewed in the narrow context of locally-produced non-motorised water lifters, they have to be assessed against buckets, handpumps, treadle pumps, animal pumps etc., while pumping itself is one option amongst many in water-supply or dry-season agriculture. The research topic has rapidly expanded to entail issues such as innovation in small enterprises, water rights, the maintenance of soil fertility, the desirability of rural industrialisation or of small-holder irrigation.

Using experimental results from England and Zimbabwe, a hydram pump was compared with two common non-motorised alternatives: the comparison is set out below. One Warwick irrigation hydram, with a delivery head equal to five times its drive head, can efficiently lift 0.35 litres per second: a typical application would be to raise 30,000 litres per day up 6 metres and across 50 mm to irrigate a 0.5-hectare plot.

Water lifting technology	Capital cost £ in 1989	Person-hours effort/day	Geographical application
Plastic hydram system (1 × 100mm)	400	0	Hill streams
Treadle pump	150	9	Shallow wells and surface waters
Buckets	5	30	Surface waters

These figures suggest that the extra capital cost of a hydram system might be recoverable within a single season.

The diagram below (Figure 1) shows one variant of a high-lift steel hydram pump employing a mixture of pipe fittings and fabricated parts. The photograph is of a plastic pump for a low-lift higher-volume application. Corrosion and cavitation are significant problems with the former, whereas fatigue failure is important for the latter.

FIGURE 1
DTU 2″ STEEL HYDRAULIC RAM PUMP (MARK 6.4)

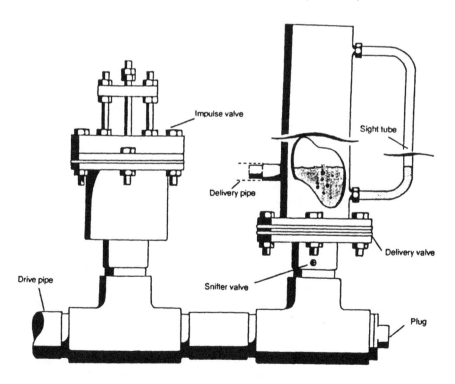

A plastic-bodied irrigation hydram by a stream in Zimbabwe. There is a 6-metre (white) drive pipe to the left of the picture and 50 metres of (black) delivery pipe to the right. In the centre, water is emerging from a long-tongued impulse valve prior to its sudden closure.

Both machines embody features based on understanding that would have been very difficult to reach in the field, away from University facilities. Now, however, that understanding must be interpreted to African practitioners, not only via a rule book ('the best drive-pipe slope is 1 in 3'), but in a way that permits future improvement after Warwick University has withdrawn from the scene. This will probably be the hardest and worst-performed part of the programme, for perhaps the hydram is too complex to be fully appropriated by its potential users.

Conclusions

The locally-manufactured hydram, in its metal and plastic variants, has still to be shown to be a viable and sustainable technology. It appears to match significant needs and its development has attained considerable momentum. Its history so far points out several lessons. The most obvious are that

- Considerable enthusiasm and perseverance are required to bring any technology development programme to a point of financial take-off.

- The contributions of both Northern specialists (here design engineers) and Southern field agents appear essential for initiating this sort of technical change.
- The informal way of working (for example, use of student projects) available to an educational establishment is quite helpful in the early stages of a new development; however, ultimately a switch to a more professional and expensive way of working is required.
- The better resources of a European university give it advantages over an African one in an area like product development. Both have diffi-culties in linking to the rural population. In the longer term, however, it is essential to locate the bulk of such design and research work in the country of final use and not in a Northern university.

In this example of the hydram pump, the integration of design for use, design for manufacture, the founding of new commercial or non-governmental organisations (for testing, manufacture and installation) and the assessment of socio-economic viability is not yet finished. The complexity of the processes, the long time and considerable resources required to develop this product illustrate why much 'appropriate tech-nology' has a rather poor track record. It is more difficult than is usually allowed for.

REFERENCES

Chambers, Robert, *Rural Development*, London: Longman, 1983.
Carruthers, Ian, *et al., Tools for Agriculture*, 3rd ed., London: IT Publications, 1985.
Terry Thomas *et al, Assessment of the Potential for Non-motorised Irrigation of Small Farms from Streams in Manicaland, Zimbabwe*, DTU Working Paper 31, Warwick University, 1989.
S.B. Watt, *Manual on the Automatic Hydraulic Ram for Pumping Water*, London: IT Publications, 1975.

New Ways in Technological Research for Africa's Less Industrialised Tropical Countries

BERNARD CHEZE

Difficulties encountered in the introduction of new techniques in developing agriculture have led to a special study of traditional farming systems and have initiated a need for a proposal for a full set of technical solutions ranging from production machinery to the processing of products – even their sale.

The need for co-ordination and for finance for research and development has shown that bilateral relationships between researchers have to be extended to a network, as in Africa through ACEMA (Euro-African Association of Agricultural Engineering Centres). Although a very wide range of research seems to be undertaken by European and African Centres, in the main it tends to be directed towards

- *reducing the cost of agricultural practices (animal-drawn implements requiring lower tractive force pull-strength, stripper harvesting of rice, low-cost threshers)*
- *fitting the size of equipment to the size of farms (one-row stripper, low-powered tractor, processing of cassava at village level)*
- *covering the extra costs of mechanisation by high added value of new types of food for urban markets (that is, new processing, such as osmotic dehydration of fruit and vegetables)*
- *developing measuring devices adapted to machines in normal field operations, for testing their suitability (on-field research)*
- *studying upstream bottlenecks (feeding of the animals, new sources of energy from biomass)*
- *and downstream ones (stabilising, drying, conditioning, packaging of products).*

This can be seen in some of what CEEMAT does.

Bernard Cheze is the Director of Ceemat, 73 rue J. F. Breton, 34000 Montpellier, France.

Introduction

Increasing food needs in LDCs up to 1990 can be satisfied only if cereal production increases faster than at present, in the face of an important growth of population (Table 1).

TABLE 1

Region	Annual Growth rate of population (1975–1990)	Annual Growth rate of population to face 1990s needs	Present annual rate of growth of cereal (1980–2000)
Asia:	2.8	3.6	2.3
North Africa and Middle East	2.5	4.9	2.3
Africa South of Sahara	2.2	4.0	1.7
Latin America		4.5	2.5

Source: Club of Sahel.

Increase in the size of cities is also a predominant trend. Between 1960 and 1985 the Sahelian population doubled, while the population of the cities grew five times greater [Colloque of Mindelo]. These people need fresh food, but also stabilised (even pre-cooked) agricultural products.

Food Analysis

These changes in consumers' habits are being studied through an integrated research programme aimed at finding out how to feed urban centres with traditional or new products, at a cost both sufficient to repay the different partners (producer, food processor, sales network) and to be able to compete with imported or easier-to-prepare products. This scheme is being developed in Northern Cameroon and Senegal.

Under the overall direction of CEEMAT, many external laboratories are involved: ENSIA (Ecole Nationale Supérieure des Industries Alimentaires), Université LYON II, Ecole des Hautes Etudes en Sciences Sociales (PARIS 1), Université PARIS X – applied economics, etc. Dynamics of food consumers in Sub-saharan cities of Africa, evaluation of technical innovations (from transfer of technology or indigenous) and analysis of the reasons form part of this study. But the main objective is to obtain a methodology able to link these 'downstream' problems with the 'upstream' production line, and justify the need for research on new equipment for agricultural production.

Need for Agricultural Machinery

Before this very recent approach, most of the decisions concerning designs of new equipment were taken from analyses restricted to the production side such as:

- potential market for a machine
- new machine for solving a precise technical bottleneck (often tending to create a new one)
- 'eureka' gliding of an inventor
- local copy to overcome foreign imports
- evolution of technology in developed countries providing major trends for the future in LDCs.

Market Study

In 1980, a study was undertaken on the need for agricultural machinery among 15 African countries.

Potential markets and a list of new machines to be designed were partly determined. But the difficulty was to make a close relation between general agricultural conditions (farming system, mechanisation policy, macro-economics) and the technological system and equipment that should prevail. Most of all, the economic situation of these countries with recent constraints imposed by the International Monetary Fund, and political choices, frequently change initial, apparently more logical solutions.

Suitable Level of Technology

Another difficulty lies in the very weak industrial network of most African countries. Lack of basic energy, raw materials, technical skill and maintenance are, up to now, some of the limitations to be found between south of Sahara and north of South Africa. M. Ogier (CINAM) has sought to explain the necessity for intermediate technology through a cross-relation between autonomy and performance of equipment: autonomy as the ratio between the number of people able to repair it and the population, and performance as a quality index from 0 to 1 (low to high).

This approach, despite being broad and non-scientific, may help towards a quick analysis of a technical choice at a local or national level.

Implementation of Various Equipments

Between 1981 and 1987, the Ivory Coast – more aware than many other countries of the reasons for the failure of past mechanised schemes – carried out tests of a wide range of methods used (from ox cultivation to tractors) on selected farms, with a follow-up by technicians and socio-economists. Promising results are now being experienced on a larger

FIGURE 1

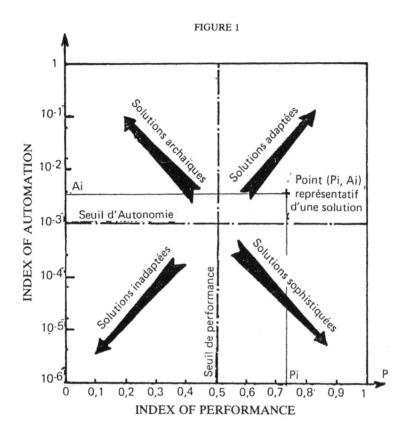

INDEX OF PERFORMANCE

scale. These are interesting but high-cost experiences. They can hardly be repeated in every country.

There is already a considerable range of available technologies. A preliminary study based on farming systems enables a selection of the more appropriate ones. An in-depth study is then necessary, on a socio-economic basis, to appreciate the acceptability of these techniques (and machines) by farmers.

Like the evolution of species in the biological system, some machines may emerge and be adopted. Very often, a lot of them will be rejected, but, even failures are useful and normal in this process.

Research work undertaken over some 20 years by CEEMAT has shown some missing 'links' on which CEEMAT and a number of other organisations, worldwide, are working. They can be classified into the following main lines of research.

Main Lines of Research

Animal Power

Used on the fringes of farming in France, animal power will remain, for the forthcoming future, the main source of power in sub-Saharan Africa. In 2000, FAO predict that 80 per cent of cultivation will be based on manual power, 15 per cent – at best – on animal power, and 5 per cent on mechanical power. Hence (despite the fact that one would imagine that all had been said on this topic, its having been used for centuries in many parts of the world), a major part of research and development has been directed towards animal power cultivation.

Less than two years ago, the Overseas Division of IER (Institute of Engineering Research – AFRC) and CEEMAT began to undertake basic research into animal traction. IER has developed a complete set of sensors to measure many physiological parameters of animals (blood pressure, rate of CO_2, heart rhythm, etc. and to try to relate these, or one of them, to fatigue under various working conditions. CEEMAT has developed new rolling types of implements that require lower pull-strength than a conventional mouldboard plough.

Energy Substitution

For countries lacking resources in oil or natural gas, biomass remains the main solution on a long-term basis, the short term being taken care of, at a low cost, by imported fuels. A small team of researchers is kept to solve peculiar problems like:

– use of vegetable oils for engines: from a small oil press with a 6 kW HATZ engine using coconut-oil, to a 600 kW diesel engine of a cotton ginning-factory modified to use directly cotton-oil produced by the same factory,
– use of ultra cleaned coal mixed with fuel oils and water (54 per cent, 15 per cent, 30 per cent) directly injected into a standard diesel engine, research conducted with CADET International and Compiegne University of Technology,
– gasifier for crop residues (after bricketting) and for rice husk.

Adapting Technology to Small-Scale Farming

Transformation of the agricultural landscape in developed countries has taken time and investment, financed indirectly by industrial develop-

FIGURE 2
ROLICULTOR

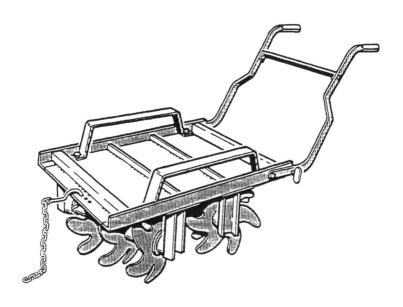

ment. Such an 'artificial' modelling of the agrarian fields was made, step by step, enabling a better use of larger and larger tractors. On the world market, technology is quickly available for tractors, at any rate, at all levels and all costs. But for rice harvesting, the choice ranges from large harvesters (with operating problems on small plots) to small and often too sophisticated Italian or Japanese binders or harvesters. A very interesting solution has been arrived at with the CAAMS'S reaper windrower and stripping system, whereby a light stripper harvester designed by CEEMAT using a 6 kW engine, and working on a 50-cm frontage, can harvest 1 ha in 6 to 8 hours. The cleaning process could be improved, but many solutions are already available on the market.

In some cases, demand leads to the mechanisation of a traditional process. For example, a complete line of machines for turning cassava into *gari* (a Togolese semolina) was designed, with a capacity of up to 500 kg/hour, enabling this to be achieved at village level, close to where the crop is grown. The final product – (minus its water-bearing roots!) packed

FIGURE 3
GENERAL LAYOUT OF MACHINE
(plan view)

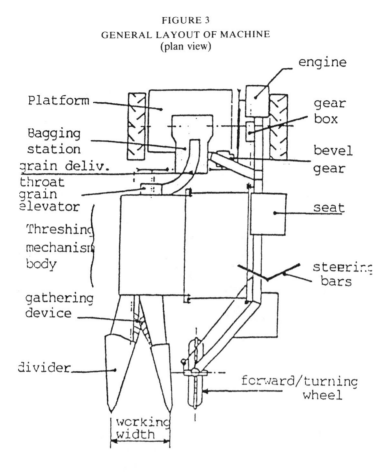

in plastic bags is sold on the urban market – even in some luxury stores in France!

Preparing New High Added Value Types of Food

Osmotic dehydration of food, a process developed by the CEEMAT food technology laboratory of Montpellier, tends to ease loss of water and limit entry of a special solution.

Solutions have a depressing water activity and remove 70 per cent of water at a low temperature 30–50°C. There is no contact with the oxygen of air. It is considered as a pre-conditioning stage that needs to be completed by stabilisation. Among the interesting facets of this system, it is possible to mention savings of six to 22 times of energy due to a transfer of liquid phase with a better thermal exchange rate and removing of water without change of state. Dehydrated fruit, vegetables and even fish or meat can be sold at a better price than fresh products.

This added value should profit the co-operative members of a village and pay back farmers with a secured price for their products. Increasing the demand in quantity and quality of local, regularly purchased agricultural products is certainly a key factor for the modernisation of agriculture – so long as the dumping of imported products does not ruin things.

Conclusion

The trend in mechanisation research for developed countries is towards the robotisation of time- and labour-consuming operations (the picking of fresh fruit, the job of milking). Expert systems – the use of microcomputers – tend also to lessen the burdens of farm managers. The technological gap seems to be still increasing between North and South. But research into new methods for tropical countries have to be chosen from this general technological progress as being suited to the existing agricultural and industrial situation in each country. No general recipe exists, nor is there any one level of mechanisation to be put forward.

Research will also bring some interesting feed-back for developing the countries. ULV spraying techniques, generated under tropical conditions (Sri Lanka), based on the use of spinning discs, are now spreading in temperate zones, not because of lack of water, but for the quality of spraying (uniformity of droplets spectra) and for reasons of economy (tanks need less refilling).

Mainly for its high reduction in power level (5 to 1), the stripping technique is also gaining interest in developed countries – already in Britain, and soon in France.

An important effort is also to be made on exchange of information between countries. At the European level, there are organisations dealing with tropical research on mechanization. At an African level, there is frequently duplication of work within very similar climatic zones. These are two of the main reasons why a Euro-African Association of Agricultural Mechanisation Centres (ACEMA) has recently been created. Institutions belonging to eight European and 25 African countries are now members of this network, the Secretariat being established in Yaoundé (Cameroon).

Solar Energy Research, Development and Demonstration in Pakistan

IFTIKHAR A. RAJA AND
MOHAMMAD G. DOUGAR

This paper presents the research and development work on solar energy carried out in Pakistan during the past few years. Various organisations and institutes involved in this work are described here, and the areas where solar application is justified are identified.

Introduction

Pakistan, having poor indigenous energy sources, and with energy consumption much greater than the supplies from its own resources, depends heavily on imported energy (Figure 1).[1] The important areas for which further energy supplies are immediately required are:

- industry;
- agriculture and mechanisation;
- irrigation;
- the basic necessities of life for the rural population.

With the exception of agricultural mechanisation, electrical energy is required to achieve the others. In Figure 2 the total electrical energy consumption is presented for the years 1981–82 to 1986–87, and it is estimated that electricity demand by the year 2000 will be 17000 MWe.[1] On the other hand, production from all available energy resources (hydro, coal, gas, etc.) would hardly reach 11000 MWe.[2] Thus, by the year 2000, the country will face a significant shortage of 6000 MWe. Such a hugh demand could by no means be met by solar energy.

Solar energy also provides an immediate solution to the energy requirements of the rural population. The rural population constitutes 70 per cent of the total, living in remote villages, where the existing supplies of

Iftikhar A. Raja is in the Department of Physics, University of Baluchistan, Quetta, Pakistan. Mohammad G. Dougar is in the Energy Studies Unit, University of Strathclyde, Glasgow. The authors are grateful to the British Council for financial support for Mr Raja's present visit to UK and to Dr J. Twidell for his encouragement to write for this conference.

FIGURE 1
ENERGY SUPPLIES AND DEMAND
(Million Tonnes of Oil Equivalent)

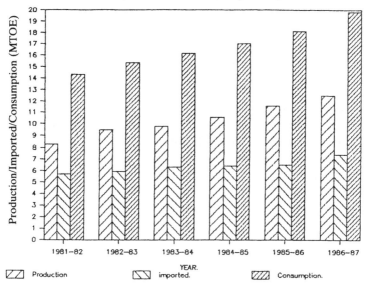

FIGURE 2
ELECTRICITY INSTALLED CAPACITY (MW)
(Pakistan)

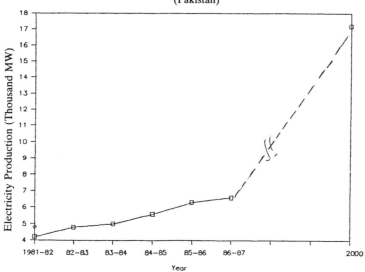

energy are non-conventional fuels, such as animal dung, wood and agriculture waste. The share of the commercial energy supply to the rural areas is less then 20 per cent.[1] The main hindrance in connecting the villages to central grid electricity is the high transmission cost. Lack of infrastructure for transportation in remote areas also restricts the use of diesel generators.

The solar energy supply systems are currently not economical and cannot compete with the available conventional and non-conventional energy supplies. However, Pakistan has started a modest programme of research, development and demonstration, so that, once the breakthrough in the technology is made, it should be in a position to quickly benefit from this natural resource. In this paper, the different agencies involved in research, development and demonstration of solar technology in the country are discussed. Also, the areas are identified where solar appliances are justified. Different steps to increase the standard and number of scientific and technical manpower are also listed.

Institutional Framework and Activities

At present the following institutions in Pakistan are involved in various aspects of solar energy applications and technology development.[3,4]

The Directorate General of New and Renewable Energy Resources (DGNER) – Ministry of Petroleum and Natural Resources.

The Directorate is responsible for the development and execution of solar energy projects. It is concentrating on decentralised solar power systems in remote areas. So far solar electrical power systems in 13 village bases have been installed with a capacity of 280 kWp[1]. The detail of installed solar systems is given in Table 1.

Energy Wing – Ministry of Planning and Development

The Energy Wing is responsible for overall planning and co-ordination in the activities of energy. A few studies on remote power generation including the utilisation of renewable energy have been commissioned by it.

Agricultural Development Bank of Pakistan (ADBP)

The ADBP lends money for development in the agriculture sector. Its main interest lies in solar power pumping systems for irrigation and it has already installed some experimental systems. ADBP evaluation through a preliminary study shows that photovoltaic applications relevant to its lending policy are not feasible. However, its assistance could be available if solar energy eventually proves to be a viable alternative to conventional technologies.

National Institute of Health (NIM) – Ministry of Health

The Institute's activities are concentrated on upgrading the health services in the rural community. The institute is interested in solar energy installations and maintenance – for example, refrigeration and lighting. NIH has installed a few photovoltaic vaccine refrigerators and is testing these in Islamabad and in the interior of Sindh.

Telephone and Telegraph Department (TTD) – Ministry of Communication

TTD is interested in remote telecommunication relay systems run by solar power. It has already installed some solar photovoltaic power stations and is successfully operating these.

National Institute of Silicon Technology (NIST) – Ministry of Science and Technology

The activities range from the development of silicon cells to testing and demonstrating the technologies of photovoltaic cells using solar collector systems, with tracking and non-tracking systems. NIST intends to develop an annual manufacturing capacity of 40 kWp of crystalline photovoltaic cells.

Universities

The activities in universities involve the basic research of solar energy utilisation. The following universities are actively engaged in solar energy research and development:

(a) University of Engineering and Technology, Lahore:
    Solar refrigeration by an absorption cycle;
(b) University of Engineering, Peshawar:
    Solar water and air heaters and low cost solar cooled houses;
(c) University of Karachi:
    Solar radiation data analysis;
(d) Islamia University, Bhawalpur:
    Solar radiation data analysis;
(e) University of Baluchistan:
    Solar and Wind Atlases for Pakistan, and solar crop drying.

Pakistan Atomic Energy Commission – Lahore

Solar lighting, heating and refrigeration systems, crop drying and water distillation.

Pakistan Council of Science and Industrial Research (PCSIR).
Ministry of Science and Technology

The work of PCSIR Laboratory, Hyderabad, includes solar water heating, solar water pumping, photovoltaic and solar desalination.

Priority Areas for Solar Power Application

The following are the areas where, at present costs, solar application is justified.

Rural Health Services

Upgrading the health services by providing solar lighting and vaccine refrigeration in the remote health clinics.

Village Electrification

About 70 per cent of the national population relies on non-conventional energy resources, animal dung, agricultural waste and kerosene for lighting, which all lead to health hazards.

Rural Domestic Water Supplies

Present water supplies are met by water collected in ponds and small streams. Over 40 per cent of illnesses are attributed to water-borne diseases.

Irrigation and Water Supplies

To acquire self-sufficiency in food production, rising fuel prices will favour utilisation of solar energy.

Agricultural Production and Storage

A considerable amount of food (grains), fruit and other agricultural products is wasted because of a lack of proper crop drying and storage facilities. Solar crop dryers and cold-stores would provide excellent means to prevent this wastage.

Telecommunication

To link the rural and urban populations.

Scientific and Technical Manpower

As Pakistan is today faced with an acute shortage of high-quality, trained technical and scientific personnel at the professional level, the following steps are worth considering:

TABLE 1
SOLAR PHOTOVOLTAIC INSTALLATION IN PAKISTAN
(BY DGNER)

Site	Capacity/Batteries	Operation Year
1. Mumnial Solar System. (Gujar Khan).	8.0 kWp/1000 AH	1981
2. Micro Padiar(Sialkot).	18.5 kWp/2700 AH	1984
3. Gakhar (Attock)	57.0 kWp/3000 AH	−
4. Dhoke Mian Jewen (Jhelum)	39.0 kWp/1700 AH	−
5. Malmari(Thattha)	10.0 kWp	−
6. Dittal Khan(Tharparker).	20.0 kWp/1500 AH.	1986
7. Khurkhera(Lasbela)	4.0 kWp	−
8. Baiker(Bugti)	5.0 kWp/ 1200 AH	1986
9. Shorozi(Kharan)	8.0 kWp	−
10. Lehtar(Kharan)	15.0 kWp	final stages status 1987
11. Mera Rehmat Khan (Abbottabad).	18.0 kWp/1500 AH	1984
12. Kankoi(Swat)	37.5 kWp	1983
13. Nasirabad	20.0 kWp/2400 AH.	−
14. Sundus	24.0 kWp/2400 AH.	−

- the creation of foundations for scholarships to generate substantial funds to supplement and enhance existing programmes for advanced training of scientific personnel;
- the development of close links with universities and research institutions of repute in developed countries;
- a Science and Technology scholarship scheme for higher studies, including renewable energy studies.

NOTES

1. *Energy Year Book 1987,* Directorate General of New and Renewable Energy Resources, Islamabad, 1987.
2. I. A. Raja and S. M. Raza, 'Nuclear Power in Pakistan', 7th Miami International Conference on Alternate Energy Resources, Dec. 1985, Miami Beach, Florida.
3. *Sixth Five Year Plan,* Government of Pakistan, Islamabad, 1983.
4. A. Eggers Lura, *Solar Energy in Developing Countries* (Oxford: Pergamon Press, 1979).

New Technology and the International Divisions of Labour: A Case Study of the Indian Software Industry

RICHARD HEEKS

This paper investigates the divisions of labour affecting the internationalisation of software production, using the example of the Indian software industry. Two international divisions of labour are seen. First, a skill division in which only the less-skilled production tasks are usually undertaken by Indian workers. Second, a locational division such that most export work takes place outside India.

A number of dynamic factors – deepening of client–developer relations; protectionist pressures; new technologies – are encouraging change. While these findings do not deny the importance of 'cheap labour', they do highlight the impact of other variables (skills, trust, risk) which make relocation of production back to developed countries unlikely.

Divisions of Labour

Under the old international division of labour, developing countries exported primary products while developed countries exported industrial manufactures. A new international division of labour was then observed in which transnational corporations sought out cheap labour sites in developing countries as 'export platforms' for manufactured goods [*Elson, 1988*]. The new international division of labour therefore rep-resents a transfer of industrial production from developed to developing countries.

Writers such as Frank [*1981*] and Rada [*1980a and 1980b*] have focused on the importance of cheap labour in promoting this new international division of labour. They argue that industrial production is transferred to developing countries because these enjoy a comparative cost advantage relative to developed countries, thanks to cheap labour. They argue further that this situation is vulnerable to new technology and other

Richard Heeks is with the Open University, Milton Keynes, England.

factors which will erode the cost advantage and produce a new pattern in which production is relocated back to developed countries.

This paper investigates the international divisions of labour within the global software industry, basing its findings on data gathered during two periods of fieldwork in India in 1988 and 1989. During these, research material was collected from appropriate government documents and computing journals. In addition, 150 interviews were conducted with software industry managers, lobbyists, representatives of trade bodies, and bureaucrats in both policy-making and policy-implementing roles, and the 30 largest software export firms (responsible for over 95 per cent of exports) were surveyed. Some additional interviews were also conducted with UK companies utilising Indian software services.

The Indian software industry is part of the new international division of labour as defined by Elson. In 1989, work worth around US$100m was contracted out to Indian software firms, mainly by transnational corporations [Government of India, 1990]. Undeniably, one of the driving forces is that Indian software labour is much cheaper than that in, say, the US or UK [Kumar, 1987], but other factors also play a part. Some of these factors are described below in relation to two additional divisions of labour seen within the overall new international division of labour. India's software exports have been characterised by both skill and locational divisions of labour.

Skill Division of Labour

Software production overall is highly skilled but is usually seen as being broken down into a series of steps, as shown below. This fragmentation of the software production process lays the ground for a division of labour, because there is a split between the earlier stages of analysis and design, which require higher levels of skill and experience, and those of coding and testing, which are relatively less skill-intensive and more labour-intensive.

This split has tended to form the basis for the international skill division of labour in the software industry. From the surveys, it emerged that the majority of software contracts awarded to Indian companies have allocated only the less-skilled coding and testing stages to Indian workers. That is to say, Indian workers have mainly been used as programmers rather than as systems analysts or designers.

This division of labour means that most software research and development work is carried out in developed countries [Rada, 1980b: 95], so consigning the Indian industry to remain a dependent follower of trends and innovations created elsewhere.

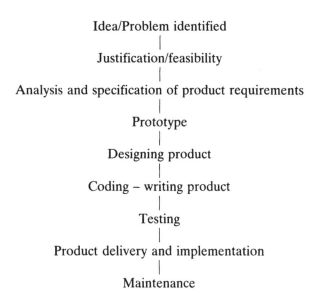

Idea/Problem identified
|
Justification/feasibility
|
Analysis and specification of product requirements
|
Prototype
|
Designing product
|
Coding – writing product
|
Testing
|
Product delivery and implementation
|
Maintenance

Factors influencing skill division of labour:

(i)   Trust and risk. According to interviewees, one principal cause of this skill division of labour is a lack of trust and a perception of risk among clients, who are uncertain of the Indian firm's skills, capabilities and credibility. In order to reduce the risk, many clients choose to retain as much control as they can over production, only contracting out the relatively unproblematic tasks of coding and testing.

(ii)   Skill profile within India. The figures below compare the break-down of software development staff in the Indian software industry, indicated by the 1989 survey, with the cross section required on a typical large turnkey project overseas (a turnkey project is one in which the contractor carries out all stages of production). Figures for the latter were provided by one of the firms surveyed.

	Indian software industry	*Average overseas project*
Project leaders	9%	14%
Analysts/designers	16%	47%
Programmers	75%	39%

It can be seen that the Indian industry is significantly 'programmer heavy'. It is best suited to providing programming labour, and the lack of analysts and designers makes it difficult for Indian companies to accept turnkey contracts.

In part, this arises because this is a relatively young industry which has not had much time to build up skills – most companies were set up in the 1980s. However, there is also a steady 'brain drain' of skilled staff, who go to live and work abroad, particularly in the US and Australia. From the surveys, one can estimate that the Indian industry loses around 15 per cent of all its export staff every year in this way.

There is also some tendency for the skill profile to be self-reinforcing. Indian company managers see that most of the export work they are offered is for programmers. They therefore feel a reduced incentive to try to raise their workforce skill profile because they see little demand for analysts and the like.

Locational Division of Labour

As well as the skill division of labour, there is also a locational division of labour, with 75 per cent of India's software export work actually taking place overseas, at the client's site, in 1988/9 according to the 1989 survey.

Factors influencing locational division of labour:

(i)   (Lack of) trust and control. As stated already, developed country companies perceive a risk in contracting out work to Indian companies. They much prefer to have software development carried out onsite where they can control the process.

(ii)   Hardware availability. Indian software companies will not always own the same computers as their clients. One of the simplest ways in which this problem is currently circumvented is by getting the Indian software developers to go abroad where they can work directly on their client's machine.

(iii)   Interaction with clients. The development of software requires continuous interaction between client and developer. In the absence of adequate communications technology, interaction takes place best face-to-face and this will mean the developers going to the client, rather than vice-versa.

(iv)   Indian company. It was stressed by many interviewees that the decisions about location of production were almost entirely based on the wishes of the client. Nevertheless, the Indian company can have some influence on the matter and may see some advantages in onsite work. It generates immediate revenue without involving much managerial or communications costs and it exposes staff to international market trends and work skills. Companies are caught in a bind because staff on onsite contracts can easily join the 'brain drain'. However, if the company prevents those staff who want to from working overseas, the staff usually just join another company which *will* send them abroad. Indian companies are therefore motivated to retain some measure of onsite work.

(v) Factors working against onsite work. Despite the advantage of onsite work to Indian companies, many of them push for more work to be done offshore, in India, because they prefer to have greater project control. On offshore contracts, the Indian firms are able to use less-skilled staff and train them up on the job; it is easier to cope if staff members leave or to add in more staff if the project is overrunning; and profit margins are generally higher. Offshore work also avoids problems with getting visas.

Clients themselves value offshore work because it is cheaper by between 30 and 50 per cent. It saves on foreign living allowances; on client managerial, administrative and hardware resources; and it avoids travel and visa delays and costs.

Conclusion

Picking up this last point, if labour and related costs were of overriding importance, one might expect the locational division to disappear quite rapidly because foreign companies save more by having work done offshore in India rather than overseas. Nevertheless, onsite working proved persistent during much of the 1980s [*Dataquest, 1989a*]. This fact and the reasons behind the skill division of labour suggest that the client–developer relationship, skill, technical and other factors are as important as labour costs in determining decisions about production.

Factors for Change

Towards the end of the 1980s, there were signs of some factors coming to bear which encouraged a change in the existing divisions of labour.

'Trust Curve'

The first few contracts that clients award to an Indian company will usually be for a relatively unimportant task which is performed onsite, and which comes with tight specifications (i.e. all the analysis and design) having been done by the client, so that there is little that can go wrong or be misinterpreted with the coding.

However, once the Indian company proves itself able to follow a set of instructions or specifications and to deliver on time and to the required quality, then it may be entrusted with a little more of the software development process and/or may be allowed to carry out more of the work offshore.

Indian software companies have therefore tended to move slowly up a 'trust curve' in terms of skills; firstly taking on only the least skilled elements of software production, then also creating the design, and finally accepting responsibility for the entire software development starting from the client's 'statement of the problem'. Similarly with location, it becomes increasingly possible that work will be sent offshore.

Change has not been great, at least in the division of skills, because it is constrained by the available skill profile, but this does indicate again the importance of trust and of the client-developer relationship in guiding the division of labour.

Visa Restrictions

Around two-thirds of India's software services exports go to the United States [*Widge, 1990*]. In the late 1980s, the US government issued tougher guidelines on work visas such that visa applications for less-skilled software workers were rejected more often than had previously been the case [*Bhagnari, 1989*]. This could have made onsite work more difficult to undertake, especially for programmers.

However, the change has been more rhetorical than real, with large numbers of Indian programmers continuing to travel to the US to work. Widge [*1990*] explains the apparent contradiction as arising from opposing political pressures. Local labour unions in the US want to halt the use of foreign workers, and this forces the US government to be seen to be doing something. On the other hand, pressure from business ensures that the door remains open because US companies are experiencing a skills shortage amounting to tens of thousands of software workers and they want the certainty of access that onsite working allows [*Widge, 1990*].

Although future trends in immigration policy are unclear, it seems likely that onsite working will be allowed to continue for some time because of the skills shortage. What is more certain is that the changes have pulled a number of Indian firms out of their complacency about onsite work. In 1989/90, several announced plans to increase offshore working.

New Technology

International telecommunication links: In the late 1980s, use of international telecommunication links by Indian software companies expanded rapidly. Companies were able to access the International Packet Switching Service (IPSS) data transmission network via an electronic 'gateway' in Bombay or to use satellite earth stations like those which Texas Instruments has been using since 1987, linking its Bangalore software development office to its headquarters in Houston [*Poe, 1987*]. By mid-1990, at least eight software export firms were using such links to access their client's mainframe computer, based overseas, from terminals based in India [*Dataquest, 1990b*], thus overcoming the problem of hardware availability in India.

Telecommunication links also enhance the ability of foreign client and Indian developer to interact on a daily basis, allowing software under development to be sent back and forth and modified according to client wishes. With greater and improved interaction, there is reduced risk and greater control for foreign clients, which encourages greater trust.

Not surprisingly, this new technology infrastructure is encouraging a greater level of Indian software production overall. By the end of the 1980s, it had created a path for the export of at least US$8m of software from India [*Dataquest, 1989a*] – exports which would not have been likely to take place without this infrastructure.

Feketekuty and Aronson [*1984*] point out that 'The international communication system serves as the central transport network of the world information economy'. It does not harm, reduce or reverse India's low labour cost comparative advantage. Instead it allows this advantage to be more manifest and accessible, making trade virtually independent of distance.

Because this new technology automates transportation rather than production, this finding does not contradict ideas about cheap labour and automation. However, it does make relocation of production back to developed countries less likely, and it reinforces the importance of factors other than labour cost.

Judging from the 1989 survey, as yet, the impact of new telecommunication links on the skill division of labour has not been great. In most cases, projects are still conceived and requirements specifications worked out overseas, so that Indian workers still only undertake the relatively less-skilled tasks of coding and testing.

However, the use of new telecommunication technology does attack many of the major problems associated with onsite working – client trust, hardware availability, and client-developer interaction – and it appears to be leading to more offshore working. By reducing the need for onsite work, such links, though expensive, can also help to save on travel, visa and living allowance costs.

4GLs and similar software tools: New software tools, especially fourth-generation programming languages (4GLs), are being introduced into software development. These act to partially automate elements of software production and this, in turn, is having an impact on the Indian software industry. Foreign clients are increasingly demanding use of these new software tools in their contracts and Indian firms are becoming more and more skilled in the use of such tools. In 1988, less than 50 per cent of the Indian firms contacted were using such tools; by 1989, the figure had risen to almost 90 per cent.

The firms benefit further from the shortage of 4GL–related skills among contract firms in developed countries [*Computing, 1987*]. Because of this, 4GL-related labour costs are rising in these countries and so India's comparative cost advantage in software production is maintained rather than reversed by these automating technologies [*Heeks, 1989*]. As a result, India's 4GL-related software exports have risen rapidly – from next to nothing in 1986 to around 20 per cent of all export contracts in 1989 according to the 1989 survey. This rather contradicts suggestions that these new technologies will discourage internationalisation of production.

The impact on the other divisions of labour is not so clear. 4GLs inspire client confidence mainly through their applicability to prototyping, in which a quick model of the final system is built early on in development. Giving clients the confidence of seeing and commenting upon a prototype system can lead to a decreased perception of risk in contracting work out to India. This could increase the amount of work done offshore, aided by the ease with which errors and misunderstandings that creep into offshore development could be corrected using 4GLs.

Similarly, 4GL portability between different computers can affect the problem of hardware availability. The bulk of development can be carried out offshore on an Indian-built microcomputer without the need to import the same type of computer as the client owns. This, again, will tend to increase the opportunities for Indian firms to carry out work offshore rather than onsite.

However, these factors appear counterbalanced by the use of prototyping which, while it might improve the client's confidence to allow work to be done offshore, in reality requires more frequent interaction between client and developer than traditional methods. Client-developer interaction of this kind tends to be synonymous with onsite software development and most of those interviewed who used prototyping undertook it at the client's site. It is not clear whether new telecommunication links offer a sufficiently interactive environment to reverse this.

A trend towards onsite work is reinforced by the shortening of the coding and testing stages with which 4GLs are associated. Because of this, a number of clients had felt that the additional organisational and communication problems of carrying out work offshore were no longer compensated by the simultaneous reduction in costs, and they had therefore decided to keep work onsite. Overall, then, 4GLs have not been associated with any increase in the level of offshore software development and they may be associated with a decrease.

As regards the skill division of labour, there are some reasons to believe that 4GLs are associated with a greater integration of the previously separated software production steps, thanks to prototyping. Because of this and their impact on client trust, 4GLs may encourage a change in the skill division of labour.

Conclusions

The findings presented here do not contradict the idea that cheap labour is a factor driving the internationalisation of production, nor the idea that export-oriented production might be relocated back to developed countries, if, for whatever reason, developing country production locations were to lose their comparative cost advantage.

However, the findings do suggest that 'Explanations of relocation based solely on cheap labour are misleading' [*Wield and Rhodes, 1988: 305*]; that comparative advantage is based on more than just labour costs; and that factors other than labour costs are helping to strengthen comparative advantage. Such factors include skills shortages in developed countries, and trust, risk and other elements of client-developer relations.

Because of the issue of skills, new production-automating technologies are not reversing India's comparative cost advantage and are therefore not encouraging a relocation of production away from this developing country. Rather, these new technologies are positively increasing the opportunities for internationalisation of production.

Other trends that do not relate to labour costs – the deepening of relations between foreign and Indian companies, the shortage of skills in developed countries, the greater use of transportation-automating technologies – are also strengthening the internationalisation of production, making relocation less likely. Nor are there any signs of multinationals actually relocating their software contracts and production away from India to other countries. Quite the opposite, in fact, with a steadily increasing level of collaboration and investment being seen [*Dataquest, 1989b; Dataquest, 1990a*]

Like other examples of internationalised production, software suffers from a skill division of labour which generally offers Indian companies only the less-skilled production tasks. More unusually, there is also a locational division of labour, with Indian software workers generally having to work at the client's site outside India. Both these divisions proved persistent during much of the 1980s, but there were signs of change towards the end of the decade.

Changes in the skill division of labour have not been particularly

marked, but are encouraged by some new software production technologies and as client-developer relations progress. There has been
stronger encouragement for the growth in offshore working, particularly
the conjunction of new telecommunication links and the threat of increasingly restrictive visa policies in the US. Greater trust between client and
developer as relationships build, and greater availability of suitable hardware within India have also helped. Once again, as with the overall
internationalisation of production, the issue of labour costs is not overriding – it is only one among a number of others.

REFERENCES

Bhagnari, S., 1989, 'Uncle Sam Gets Tough', *Dataquest*, Feb. 1989, pp. 50–53.
Computing, 1989, 'Quickbuild Staff Get Rich Quick', *Computing*, 5/11/87, p. 2.
Dataquest, 1989a, 'The DQ Top 20 1988/9', *Dataquest*, July 1989, pp. 57–157.
Dataquest, 1989b, 'Data Bits', *Dataquest*, Oct. 1989, p. 49.
Dataquest, 1990a, 'Seepz Exports up 56%', *Dataquest*, May 1990, p. 32.
Dataquest, 1990b, 'The DQ Top 20', *Dataquest*, July 1990, pp. 39–177.
Elson, D., 1988, 'Transnational Corporations in the New International Division of Labour:
    A Critique of the "Cheap Labour" Hypothesis', *Manchester Papers on Development*,
    IV(3), 352–76.
Feketekuty, G. and Aronson, J. D., 1984, 'Meeting the Challenges of the World Informa-
    tion Economy', *The World Economy*, 7(1), 63–86.
Frank, A. G., 1981, 'Crisis in the Third World', London: Heinemann.
Government of India, 1990, 'Annual Report 1989–90', Department of Electronics, Govern-
    ment of India, New Delhi.
Heeks, R. B., 1989, 'New Technology and the International Division of Labour: A Case
    Study of the Indian Software Industry', DPP Working Paper No. 17, Open University,
    Milton Keynes.
Kumar, A., 1987, 'Software Policy: Where Are We Headed?' *Economic & Political
    Weekly*, 22(7), 14/2/87, pp. 290–94.
Poe, R., 1987, 'India's Soft Hopes', *Datamation*, 1/9/87, pp. 96.5–96.12.
Rada, J., 1980a, 'Microelectronics, Information Technology and Its Effects on Developing
    Countries', in J. Berting, S. C. Mills, H. Wintersberger (eds.), *The Socio-economic
    Impact of Microelectronics*, pp. 101–46, Oxford: Pergamon Press.
Rada, J., 1980b, 'The Impact of Microelectronics', Geneva: ILO.
Widge, N., 1990, 'Uncle Sam Wants You', *Dataquest*, March 1990, pp. 74–93.
Wield, D. and Rhodes, E., 1988, 'Divisions of Labour or Labour Divided?', in *Survival and
    Change in the Third World*, B. Crow and M. Thorpe (eds.), Cambridge: Polity Press,
    pp. 288–309.

The Three Decades of Africa's Science and Technology Policy Development

JOHN W. FORJE

This paper traces the trends in the development of science and technology policy in Africa, from its emergence as a formal concern of governments in the early years of independence to the present day. It explores some of the possible political, economic and social implications of science and technology in the development process.

The conclusion is that the generation and utilisation of indigenous technology remains a prime factor for attaining self-reliant and sustainable development, and, independence.

Introduction

With all its human and natural resources, poverty, underdevelopment and dependency loom over the African continent. Population continues to increase at the current rate of 3.1 per cent per year – the highest in the world. This population growth rate offsets the 1.5–2 per cent per capita growth rate in the GNPs of the countries of Africa. Over 40 per cent of the continent is affected by drought. About 50 per cent of the land area of the continent is classified as arable, permanent meadows, pastures, forest and woodlands. Of the 16 per cent considered suited for agriculture, only 7 per cent is actually cultivated.

The wealth of the continent is unevenly distributed among the 51 independent nation-states. Some are poorly endowed with natural resources, some are constant victims of natural hazards such as cyclones and floods, and some are situated in poorly irrigated areas.

Given the varied nature of the socioeconomic potentials and development priorities, the pursuance of science and technology policy differs

John W. Forje (Fil dr) is a Research Fellow at the Institute of Human Sciences, Ministry of Higher Education, Computer Services and Scientific Research (MESIRES), Yaounde – Cameroon.

TABLE 1

PERIODS IN AFRICA'S S&T POLICY DEVELOPMENT

Period	Assumptions	Characteristics
1960–1970	Science as a vehicle of progress	Era of development assistance. Formulation stage.
1970–1980	Science as problem solver	Period of crisis: famine debts, energy, etc. Implementation phase.
1980–1990	Science as a source of strategic opportunity	Structural adjustment era. Evaluation.

from country to country. However, there are some identical factors and problems common to these countries which makes it possible to generalise about their policy approaches.

For simplicity, the three decades of Africa's science and technology (S&T) policy development have been divided and are looked at here as per Table 1.

1960–70: Science as a Vehicle of Progress

Science policy as a tool for progress acquired its significance in the first decade of Africa's independence. During this period national and international efforts were generated towards the establishment of a central and authoritative mechanism for the formulation and growth of science policy and its relationship with the various other policies of national development.

The formulation, coordination and even application of science and technology was, and continues to be, reinforced by the strength and weakness of the scientific traditions transplanted during the colonial period, by the particular dynamics of the transition from colonial science to science assistance, and by the variants of scientific nationalism embedded in dependent circumstances of Anglophone and Francophone African countries.[1]

Africa's science and technology policy started from the basis of dependency, an alienation from anything indigenous and motivated by the quest for rapid industrialisation at all costs. Aid assistance and industrialisation addressed only the needs of the affluent. Science policy formulation and application bypassed, from the very beginning, the acute needs of the marginalised majority.

Despite the dependency attributes of exogenous science and technology, however, the efforts of governments on the overall scientific and technological activities in the region, though slow, have none the less made some progress from the humble beginnings of the 1960s. The growing awareness within government circles of the role of science and technology in socio-economic development was significantly stimulated by the international community (the UN and its related agencies and donor nations) through arranging meetings and sponsoring missions to help formulate guidelines for the creation and growth of science policy-making organs.[2]

Thus the Lagos Conference (1964) set out to articulate solutions to the planning, organisation and financial outlay for science and technology activities; the Yaounde meeting (1967) focused attention on evaluating the state of science policy and research organisations, whilst the Addis Ababa symposium (1970) drew up an action plan for the continent.[2] The Lagos Conference was handicapped by the absence of a concerted national science and technology policy, and the lack of a national machinery for formulating and co-ordinating such policies; the Yaounde meeting revealed significant developments by African countries in establishing and reinforcing their national agencies for science planning, decision-making and co-ordination. The Addis Ababa symposium went further in laying the foundation for the First Regional Conference of African Ministers responsible for the development and application of science and technology to development in 1974.[3]

Although science and technology policy at the end of the 1960s was still in an evolutionary or tentative stage in most African states, there was a broad concensus across the continent on the future role of science and technology in the socio-economic transformation of the region. The general agreement being that science and technology, and research and development embodied the symbolism for progress, hence the urgent need for the scientific and technological upbuilding of African countries. Within the context of actual science policy development in the 1960s, one finds, in spite of the colonial imprint, some understanding of the 'needs of science' and of the 'dynamics of scientific growth'.[4]

However, the lack of a scientific community and the dominant role of the governments dictating the pace and tempo in the articulation of science and technology policy prevented any form of thorough public debate on the issue. This absence of academic exercise in a way deprived the continent from setting up the appropriate structures for scientific activities right from its inception and the institutions established failed from the very beginning to critically examine their role and functions and the purpose of their existence.

The 1960s remained, by and large, a period of discussion. In the first phase of Africa's science policy, thinking was largely guided by the same ideals which underpinned the evolution of science policy in the colonial period.

1970–80: Science as a Problem Solver

In considering the evolution of the 1970s, we must again be cautious, yet vigilant, concerning the unfolding developments that took place.

During this period basic human needs were not provided for and even the existing living standards were declining fast as a result of the magnitude and dimension of the series of crises (famine, drought, debts, political instability, etc) which rendered the nature of the transition difficult. Economic growth ceased to be regarded as an unmixed blessing, and the attainment of basic needs remained remote from the marginalised poor. Africa pathetically 'entered a second era in the government research and development (R&D) relationship, one governed by the belief that science and technology could be mobilised by government to directly solve urgent national problems. R&D was put forward as a means of solving one problem after the other; housing, the crisis of the cities, health, food, water, environment, and ultimately the energy crisis'.[5]

Whilst government efforts continued in furthering the formulation of science policy organs through national and international activities/participation[6] there was much confusion about deteriorating world prices for export crops, the exorbitant cost of manufactured products, famine and drought. All this made it impossible for African governments to meet their science and technology objectives. On top of this the international community developed cold feet on its aid assistance programme to developing countries. As a consequence a sense of disenchantment with science and technology *per se* seemed to gain ground with the public, who, seeing a rapid fall in living standards, started to question all assumptions about the dynamics of scientific growth.

In spite of such questions, hidden somewhere was a general belief that science and technology possessed all the ingredients for performing the magic trick of providing solutions to underdevelopment in Africa. The most significant development of the 1970s was the first Conference of the ministers of African member states responsible for the Application of Science and Technology to Development in Africa (CASTAFRICA),

Dakar, Senegal, in January 1974. In effect, the outstanding achievement of CASTAFRICA 1 was perhaps its ingenuity in increasing the region's awareness of the primary role of science and technology in the overall development of a country. This awareness, strengthened by international cooperation, led to the initiation of a number of efforts aimed at creating national scientific and technological capabilities.

The process of structural changes, quasi though they were, could be seen through the number of missions or assistance by the UN and its agencies (Unesco, UNDP, WHO, etc) to African countries; these sought to set out the guidelines for national science policy organs.

As a result of these structural changes, one witnessed the emergence of either a Ministry of Higher Education and Scientific Research, a General Delegation for Scientific and Technical Research, a Secretariat for Science and Technology, or a Council for Scientific and Industrial Research attached either to the Office of the President or Prime Minister with more executive powers.

For all that, African countries failed to give science the necessary push as a problem solver. The basic reasons for this were (a) poor integration of science and technology into national development plans, (b) the non-familiarity of technology policy-makers with the methods for planning and programming the activities of the scientific enterprise and of integrating these activities into an overall national development plan(s), and (c) the dependence on foreign research institutions as outposts for domestic research (very often the activities of these foreign-operated research bodies do not directly relate to the country's socio-economic objectives, and, in some cases, national policy-makers have virtually no say in the design and control of such research programmes).[7]

Two significant trends can be seen in the effort to cement the role of scientific enterprise as problem solver: (i) low-income countries, despite the economic crises, were exposed to the realities of meeting the essential needs of the people; and (ii) a determined effort was made to increase agricultural output and other activities within the processing sector, through increases in the number of labour-intensive small- and medium-size industries.

Looking at the content of science and technology policy, criteria and priorities of the 1970s, African states were slowly moving towards the direction of articulating problem-oriented science and technology policy structures. Given the evolving international climate of the 1970s, science and technology policy in Africa came closer to the growing concern and assumption of its being a problem solver by ensuring the relevance of the nation's input mechanisms in the scientific enterprise and directing these towards a variety of economic and social goals.

1980–90: Science as a Source of Strategic Opportunity

Science and technology policy entered the 1980s from a different perspective, perhaps radically different, though in continuity with the developments of the past two decades. The thrust to this radical change could be attributed to the acceptance, consequences and results of the continued state of scientific and technological backwardness and the general level of socio-economic development of the region since the attainment of political independence in the 1960s.

Emerging new socio-economic problems (nationally and internationally) considered to be structural, and with major implications for science and technology policy, could no longer be pushed aside. The problems were mounting and solutions to these problems continued to be non-existent – creating uncertainty and insecurity to every household. A bleak future hung over the continent.

Some of the developments which characterise the present situation include: a persistent combination of the impact of numerous crises; issues of environment and sustainable development, and depletion of natural resources; the emergence of new social aspirations and values; a growing scientific and technological gap between the industrial and non-industrial African countries; and unstable commodity prices and the rapid decline in the purchasing power of the African countries.

The 1980s signalled Africa's entry into the structural adjustment decade, with science and technology gaining more and more significance as a source of strategic opportunity for revamping ailing economies. An important input in this direction was the formulation of the Lagos Plan of Action and Africa's Priority Programme for Economic Recovery, 1986–1990, adopted by the conference of Heads of States and Governments of the Organisation of African Unity (0AU) in April 1980 and July 1985 respectively. Also worth mentioning is the United Nations special session on the economic problems of the continent. Various workshops on the harmonisation of science and technology policies, and the financing of co-operative R&D activities and research projects gave added momentum and dimensions to science as a source of strategic opportunity.[8]

The Lagos Plan of Action (LPA) appeals to African states to invest sufficient resources in the promotion of science and technology, and within the realms of the need to remain rooted in one's home ground, though becoming receptive to external influences and developments; that is, outside inputs should only be a complementary component to indigenous efforts, thereby reinforcing recommendations of CASTAFRICA 1 (1974). (Recommendation No. 4 appeals to all African states to establish the appropriate machinery to be responsible for the generation of a national science policy and capable of promoting genuine research for development.) Furthermore, science and technology are among the sectors given priority in Africa's Priority Programme for Economic Recovery 1986–90.

As the decade characterised by structural adjustment programmes rolls on, understanding the connection between research and production and the proper maximisation of resources, through either the concept of a 'science-push' or 'demand pull', provides the appropriate basis on which to understand the problems and to seek solutions.

Science as a source of strategic opportunity becomes even more significant in view of the possible development of new strategic industries based on biotechnology, information science, electronics, robotics and new materials. Genetic engineering/ biotechnology will also determine the place of natural products on which most African countries depend for their export earning capacity to solve outstanding financial and other developmental problems. An Advance Technology Alert System for Africa is therefore imperative at this stage for institutes and industries to maximise potential applications of emerging or frontier technologies.[9]

The Future

Africa's science and technology policy in the 1990s should be radically revitalised to meet the changing needs of society since the approaches of the previous decades have failed to provide basic human needs. From being a vehicle of progress (science push) to a problem solver (demand push) and as a source of strategic opportunity, science and technology has entered a complex period, in which emerging new technologies threaten the traditional export-oriented agricultural product base of African countries. The 1980s having been an era of 'strategic opportunity', African countries should redouble their efforts in building the necessary basic organs and structures for a flourishing scientific enterprise. At present, the continent is going through a more complex period of crucial understanding of the role of science and technology in the development process. It can be argued that the earlier periods were necessary to arrive at the present era.

Generally, science and technology policy thinking in Africa has developed in line with the assumptions advanced. Institutional reform has been slow, though there has been growing awareness of the need for science and technology in nation building. However, what have been the implications for the further development of science policy in Africa? Here we discover that broad consultation, networking and co-operation among African countries and researchers have intensified over the years.

How have these networks and co-operative ventures fared during the period? To a large extent, they have paid off in many aspects; otherwise the region could not have arrived at its present stage, even though one would have expected output.

The complexity of scientific and technological strategy in system terms corresponds to the needs and actual potentials of a country. The needs of the African people are many, given the level of under-development, external and internal exploitation of the marginalised, and the dependency of the continent; hence the urgency for African countries to work towards attaining a high level of maturity and self-reliance. The social component of science and technology and social organisation and administration of scientific and technological developments should be paid due regard in future policy considerations. This is important in view of the scale and complexity of the scientific and technological systems which are being established and which will have clear repercussions for society.

NOTES

The author would like to express his thanks to the British Council, Yaounde, for sponsoring his participation in the Conference, and to the Ministry of Higher Education and Scientific Research and the Director of the Institute of Human Sciences, Yaounde, for permitting him to attend. The hospitality of the organisers of the Conference is greatly appreciated.

The views expressed here do not necessarily reflect the opinion of the British Council, the Ministry or the Institute.

1. T. Eisemon, C. Davis and E. M. Rathgeber, 'Transplantation of Science to Anglophone and Francophone Africa', in *Science and Public Policy*, Vol.12 (1985), pp. 191–202.
2. See the following:
   – United Nations Conference on the Application of Science and Technology to the Development of Less Developed Countries, Geneva–Switzerland, 1963;
   – The International Conference on the Organisation of Research and Training in Africa in Relation to the Study, Conservation and Utilisation of Natural Resources, 28 July–6 August 1964, Lagos, Nigeria;
   – Symposium on Science Policy and Research Administration in Africa, 10–21 July 1967, Yaounde – Cameroon.
   – The Regional Symposium on the Utilisation of Science and Technology for Development in Africa, 5–16 October 1970, Addis Ababa, Ethiopia..
3. See following:
   – Unesco (1964), 'Final Report of the Lagos Conference', 28 July–6 August 1964, Ref. N.S.64/D/36.A, Paris:
   – Unesco Document SC/WS/428, 'Unesco Field Science Office for Africa, Kenya', Doc. SC/CASTAFRICA/REF/4.
4. John W. Forje, *Science and Technology in Africa* (London: Longman, 1989).
5. Roland Schmitt, 'Continuity and Change in the US Research Systems in Current

Science and Technology Policy Issues', Occasional Paper No. 1, School of Public and International Affairs, George Washington University, Washington, DC, 1985.

6. See UN Mega Conferences of the 1970s; Environment (1972); Population (1974); Food (1974); Women (1975); Habitat (1977); TCDC (1978); and UNCSTD (1979).

7. See Unesco, 'Science, Technology and Endogenous Development in Africa; Trends, Problems and Prospects', SC–87/CASTAFRICA 11/3, Paris, 1987, pp. 13–14.

8. See the following workshops, conferences and meetings:
    – Workshop on the harmonisation of science and technology policies and on the identification and financing of co-operative R&D in West Africa, (Dakar–Senegal, May 1979) and on sub-regional co-operative research projects in Eastern and Southern Africa (Nairobi–Kenya, December 1980).
    – Symposia on the functions and effectiveness of national science and technology policy-making bodies in the countries of West Africa (Lomé–Togo, May 1982); in the countries of Eastern and Southern Africa (Nairobi–Kenya, March 1985) and in the countries of Central Africa (Bangui–Central African Republic, September 1986).
    – Meetings of experts to review the follow-up of CASTAFRICA 1 (Nairobi–Kenya, October 1983).
    – Meetings of national science and technology policy-making bodies in the countries of intertropical Africa (Dakar–Senegal, July 1985).
    – Conference of Ministers of Education and those responsible for Economic Planning in African Member States (Harare–Zimbabwe, June–July 1982).
    – Conference of African Governmental Experts on Technical Cooperation among African Countries (Nairobi–Kenya, May 1980).
    – Seminar on Technology policies in the Arab States (Paris–France, December 1981).
    – Meeting on the strengthening of scientific and technological capabilities of African countries (Brazzaville–Congo, November 1982).
    – Meeting of the First Congress of African Scientists (Brazzaville–Congo, June 1987).
    – Joint meeting of International Coooperation for African Technological Development (Dakar–Senegal, December 1983).
    – OAU expert meeting on the implications of new technologies for development in Africa (Mbabana–Swaziland, November 1984).
    – Numerous conferences and Unesco/UNDP etc. missions across the continent on the preparation of CASTAFRICA 11, and national S&T activities in various countries.
    – The Second Conference of African Ministers Responsible for the application of Science and Technology to Development in Africa CASTAFRICA 11 (Arusha–Tanzania, 6–15 July 1987).

9. See Atul Wad and Michael Radnor, 'Frontier Technologies in African Development', Working Document prepared for the Joint Meeting on International Cooperation for African Technological Development, organised by UNCSTD (Dakar–Senegal, 5–7 December 1983).

MANPOWER DEVELOPMENT

Technical Manpower Development with Examples from Three Countries

ANDERS NÄRMAN

African industrialisation is at a low level, compared with the rest of the world. Among other factors, it is hampered by a shortage in technical manpower. In this paper an attempt is made to evaluate some various educational reforms, from a broader labour market and general development concept. Too often assessments of technical/practical schooling are based on a narrow economic perspective.

This paper argues the need for technical diversification at school, as well as improved vocationalised training. However, this can only be an addition to a clear strategy on general industrial development.

Background

Africa is still by far the least industrialised continent. Industrial development is hampered by numerous barriers, which can be related to economic, administrative and political factors. One obvious shortcoming is the lack of technical manpower, both on a middle, as well as a more advanced level. Too often technical development is managed by foreign experts and volunteers. A large part of the development assistance granted is to be found within this very field. Naturally this is one of the reasons why Africa remains structurally dependent on the industrialised nations of the world.

In view of the high rate of unemployment, the technical manpower shortage could be expected to be a matter of education/training. Since the early 1960s a call for a more relevant educational system has been heard from the newly independent African states. We can also register many attempts to rectify the present state of affairs. It has been debated

Anders Närman is with the Department of Development Studies, University of Gothenberg, Gothenberg, Sweden.

whether schools are the right medium to transfer technical skills on a broader scale. Opponents claim that school training is too costly, that schools do not have the ability to influence opinions, or technical subjects would only be regarded by students as a substitute for purely academic ones. However, there are few empirical studies on which to base this negative view on the potential for schools to contribute to a better general technical knowledge among the African populations.

This paper will give a brief account of three different modes of technical schooling/training, from Kenya, Tanzania and Botswana. Kenya is chosen as an example of diversified secondary school education. For Tanzania one particular vocational training centre is referred to. The Botswana example is of brigade training – education with production. All three cases have been funded by external assistance, primarily from the Swedish International Development Authority (SIDA).

The data presented here are based on tracer studies carried out by the author, sometimes in collaboration with research students. For more detailed accounts of any of the projects see the notes at the end of this paper.

Diversified Secondary Education in Kenya[1]

Compared with its neighbours, the Kenyan economy is fairly diversified. Partly this is due to the fact that Kenya has established an open door policy towards foreign investment. This in turn can be traced back to a policy of consistency with past colonial trends which was adopted at the time of independence. At that time the Kenyan economy was already fairly advanced commercially, with a backbone in large-scale cash crop agriculture.

Although Kenya has experienced comparative economic prosperity, the growth in labour market opportunities has not followed suit. For a working-age population of some 10 to 11 million (1989), a mere 1.4 million in the wage-earning sector must be regarded as a moderate achievement. The expansion of the labour market has been slightly ahead of population growth.

In spite of the fact that Kenyan economic fortunes have been connected to a substantial growth in industrial production, this has not had any greater impact on employment within this field. Manufacturing industries have provided employment for 11 to 13 per cent of the total labour force since independence. This can largely be seen as a reflection of the capital intensive character of much foreign industrial investment. Kenya might have gained from an open door policy towards foreign investments in macro-economic terms. However, very few positive effects have filtered down to the broad mass of the population, for example in industrial wage-earning vacancies. Wage earning opportunities are mostly opened up within what can be broadly classified as the service sector.

Whilst Kenya has experienced a fairly moderate growth in the labour market since independence, education has seen a tremendous expansion. A disproportionate increased demand for labour market vacancies was therefore to have been expected. (The improvement in the educational system of Kenya is primarily of a quantitative character. On the other hand, Kenya has been devoid of any reforms that would substantially raise the quality of the education offered. At the secondary level the massive intake has even depleted the previous standard consider-ably.) The number of school leavers seeking employment is at present far above new available positions for work. Not only are primary school leavers far in excess of employment vacancies, students from secondary school and even university find ever-increasing difficulty in securing gainful employment. Kenya is faced with an extremely severe school-leaver unemployment problem.

Although the formal school system is basically of an 'academic' character, there is a wide variety of technical education/training offered. School-based technical education has long traditions in Kenya. At the time of independence there were some so-called trade schools and a subject called industrial arts was offered at some schools. Since the end of the 1960s the development of Industrial Education (IE) and Technical Secondary Schools (TSS) has been largely sponsored from development assistance funds. In both cases SIDA has been involved. IE is either offered as a combination of woodwork and metalwork, or electricity/power mechanics, and is intended to be a pre-vocational subject. It takes up only a minor part of the timetable (six recommended periods out of 40 per week). Like IE, the practical subjects at TSS are of a pre-vocational nature, but are given greater weight in the timetable. At Form III–IV half the teaching time is devoted to practical specialisations. Some of the TSS are offering building-oriented subjects, such as plumbing, carpentry, masonry, or general engineering, motor vehicle mechanics, electricity, etc. Both the IE schools and the TSS are of high standard. All of them are fully government maintained, which to a certain extent vouches for good quality. In the secondary school examinations one of the IE subjects can be included. Students from TSS have both theory and practical for the special trade included in the examinations, together with technical drawing.

One common assumption about technical diversification in an otherwise academic curriculum is that it is normally taken as a second option. Results from two separate tracer studies on IE and TSS respectively give a different picture. The often-repeated 'myth', that technical education is not demanded among students in a country like Kenya, seems to be invalid. Technical secondary education, or components of it, seem to be an attractive opportunity for many. This is clearly indicated in answers given to questionnaires, as well as during discussions with students. Furthermore, the academic standard of students with IE in the examinations or the TSS ones is not in any way inferior to their peers. Neither IE, as an alternative subject, nor the TSS are shunned by the best secondary school students. These options can hardly be regarded as some kind of second choice.

During our interviews it was obvious that many former students of IE, or from the TSS, regarded the technical skills obtained as being useful. In most cases they could mention many private activities for which the skills acquired had been valuable. However, only a minority had been able to secure a job based on their particular skills. In view of the short time perspective, between the examinations and our interviews, nothing else could actually have been expected.

Obviously there is a great difficulty in entering the labour market at all. Compared to the number of Form IV school leavers, few new-wage earning vacancies are established. Among our IE sample a high share continued their schooling in Form V. A few others had taken up some kind of directly vocationalised technical training. For ex-students in training, with IE, some kind of technical bias is distinctly higher, compared to the 'control' sample interviewed. For those having been to TSS there is an even higher tendency to opt for some kind of technical training.

On attitudes to work, there is a clear correlation between exposure to a technical specialisation and an aspiration to work within that sphere. It seems too, that a pre-vocational course gives a greater degree of confidence to succeed in a preferred career. On actual jobs held after school we have to take account not only of the overall difficult labour market conditions, but also the high rate of those entering continuous schooling/ training. However, students with IE among the 15 schools offering this subject included in our sample are, to a lesser degree, totally without gainful employment. Bearing the low numbers in mind, it can be noted that IE students and those from TSS are in most cases accepted for work with some kind of technical skills needed. This is except for the most common work entered into, i.e. unqualified teacher. Even here, though, many former IE students are specifically enrolled to teach some kind of a practical subject, for example, woodwork.

It is difficult to discuss here any kind of labour market advantages from technical specialisation, based on the data referred to. It has not been possible to follow up the samples over a longer period. However, within an otherwise stereotyped curriculum, technical skills acquired can be the positive factor that will secure some kind of employment, permanently or temporarily. It might also be effected by a positive attitude developed towards any kind of manual labour. Furthermore, the skills can be used for different kinds of self-employment, especially in the rural areas, outside a formal work market.

In conclusion, it could be argued that, with a development policy like the Kenyan one, there is a demand for a higher general level of technological understanding. This can be provided by, among other methods, a technically diversified secondary education.

Vocationalised Training in Tanzania[2]

In spite of certain basic characteristics being similar for Kenya and Tanzania, the first couple of decades after independence show a totally different pattern of development. Tanzania has to a large extent followed its own populist brand of African 'socialism'. From a purely economic point of view, the development strategy chosen has not been successful, at least not so far. However, this is not to deny certain positive humanitarian aspects of the Tanzanian development, lacking in Kenya.

As in most African countries, agriculture is the dominant economic sector. The 1978 population census reveals that at that time some 80 per cent of those engaged in economic activity earned their living from agriculture. Only a minority (approximately 7 per cent) were involved in some kind of waged employment. There has been no noticeable proportionate increase in the formal labour market during the 1980s. In 1987 there was an official workforce of some 700,000 employees.

As in Kenya, services of one kind or another form the most important employment sector. The services sector also offers the greatest growth in career opportunity. The manufacturing sector has experienced a slightly fluctuating trend during the last decade. To a large extent the industrial sector is dominated by parastatal companies. Efficiency is low among Tanzanian manufacturing industries. The construction sector has experienced both a proportionate and absolute decline in terms of employment, which in itself is an indicator of general economic hardship.

Among the problems of the industrial sector, referred to in a survey carried out in the Kilimanjaro region (where Moshi is located) were those primarily related to general economic decay. Availability of spare parts and raw materials was commonly brought out in the discussions, as well as the lack of financial means to acquire the inputs. Another factor was the extremely poor infrastructure (communications, water and electricity) network. A more blurred picture emerged of the demand for skilled manpower. Some industrialists agreed that this was a problem, but they did not raise it themselves. There were even claims to the effect that would indicate a fairly good supply of skilled workers. In a way, this view can be countered by numerous complaints of poaching for technically trained employees among industries.

Tanzania, like Kenya, offers a wide variety of technical education/training. Secondary education, for example, offers a few special diversifications, one of them being the technical bias. There are also a number of development colleges. Vocational training is regulated by a Vocational Training Act of 1974. At the end of the 1970s an NVTC was set up in Changombe, outside Dar es Salaam. Thereafter another 5 NVTCs were established, offering almost 35 different trades.

The Moshi NVTC was opened in 1983. It was established as part of SIDA development assistance towards Tanzanian vocational training. In all, 13 trades are offered at Moshi NVTC. Some of them are of a more advanced character, and need some kind of a technical background as an entry requirement. For others, only a primary school leaving certificate is necessary. Most of the trainees undergo a one-year institution-based course. In some trades, all, or some selected, trainees stay on for two years at Moshi NVTC. However, the total training is intended to last for four years, with the remaining two or three years being spent at some industrial establishments for in-plant training.

A tracer study of Moshi NVTC trainees was initiated in 1986, whereby an attempt is being made to follow trainees at the centre, through in-plant training and into the labour market. As this is an on-going project only some preliminary data can be discussed here.

Many of the ex-trainees from the first two groups were able to secure employment after completing training. Especially in the case of the very first group, many were so-called sponsored trainees, i.e. the training was paid for by an employer.

The work of the ex-trainees bears at least some kind of relation to the tasks carried out and the training undertaken at the centre.

Obviously, the training at Moshi NVTC provides a suitable background for employment in some kind of technical sector. This must be regarded as quite positive in relation to the small labour market available, referred to above. However, as the recruitment base has changed, since the two first groups, both in respect of previous schooling and work experience, it will be important to examine whether the positive trend persists.

There are some aspects that have to be looked into closely if the positive trend is to be maintained. In-plant training must be regarded as an integrated part of the total course. Many trainees are never placed for this period at all. For some this part is irrelevant to the institutionalised training. Furthermore, it is obvious that neither industrialists nor the staff at Moshi NVTC are clear about the training aspect of this apprenticeship period.

The larger industrial enterprises in Moshi do have their own training workshops and instructors. It might be a better utilisation of avail-able resources to open up a more flexible integration of institution-based and industrial training. For this, channels of communication between Moshi NVTC and the industrial environment must be strengthened.

One parallel between Kenya and Tanzania is the fairly small number of work opportunities available in some kind of technical/industrial sector. We have referred here to data extracted from a survey on vocational training at the Moshi NVTC. For the first couple of years it seems that trainees are absorbed in relevant work after a completed course. It will be important to find out how the trend is maintained, with a somewhat changed recruitment base. Furthermore, some of the training objectives have to be adhered to more closely. This applies particularly to directives concerning in-plant training.

Education with Production in Botswana[3]

Compared with Kenya and Tanzania, economic development in Bots-wana is quite different. Botswana is one of the most expanding economies in the world. This is due in particular to a favourable economic situation as regards deposits of raw materials, particularly diamonds. Positive *economic* trends aside, it is also important to observe the country's continuing dependence in relation to neighbouring South Africa. Further-more, aggregate economic development is not reflected in the distribu-tion of advancement.

Botswana is a sparsely populated country with only some 1.3 million inhabitants. Some 550,000 are to be seen as potential members of the work force. In 1984 there were some 110,000 people within the formal wage-earning sector. During the previous decade, work opportunities increased by an average of 7 per cent annually. In the context of extreme economic growth, construction forms an important field of employment, in which no less than 16 per cent of all employees are engaged. This is in clear contrast to the situation in Kenya and Tanzania, as given above. Manufacturing industries employ approximately 10,000 wage-earners, that is, 13 per cent of the formal wage-earning sector. During the last decade the number of jobs in the manufacturing sector doubled. Ownership within the manufacturing sector is predominantly foreign.

Even if economic growth has not benefited the majority of the population through improved salaried working opportunities, other social advantages have pertained, such as improved educational standards. Apart from 'academic' formal schooling there is a wide variety of voca-tional training. Some of the technical institutions are Botswana Poly-technics and the Automotive Trades Training School. A number of Vocational Training Centers (VTCs) have also recently been built.

One special category of vocational training in Botswana is education with production, the so-called brigades. The initiative for these was taken some 25 years ago by a South African refugee, by the name of Patrick van Rensburg. One of the basic principles for the brigades was self-reliance, established in a combination with training and production. The productive component was also aimed at instilling a positive attitude towards practical work. Brigades were to be an integral part of the local community. There was to be distinct autonomy in the governing of the individual units, with active participation by the students themselves.

Over the years, the brigade movement has passed through different stages. Many critical situations – economic, administrative and political – have arisen. From a current standpoint, the brigades are not exactly what they were envisaged 25 years ago by their founders. Brigades today do have a fairly firm structure, co-ordinated by the Ministry of Education. Many development assistance donors provide funds for different aspects of brigade development and maintenance. Therefore, the need to combine the training with production is not a necessary prerequisite, at least not for financial reasons.

In 1988 a pilot tracer study of the brigades was initiated. For this survey four brigades were chosen. Trades offered at these brigades were building, carpentry, plumbing, electricity, welding and mechanics, i.e. mainly building ones. At the time of the survey we found 20 per cent unemployed. Almost all of the others were in employment, with a few in some kind of further education/training. Almost all the employed were working within the trade they had been trained for. From this it would appear that the brigade training has been fairly successful in job preparation, during the last few years at least. However, we have to evaluate the effect in relation to the very positive labour market trends. Many of the ex-trainees are to be found, as noted above, in building trades, for which the construction boom has immense needs.

To evaluate the brigades as a model for education with production is a more complex task. Production is not a main source of income for many brigades. Furthermore, they are hardly the self-reliant institutions, geared towards rural development, that were originally intended.

Concluding Analysis

This paper is based on tracer studies data from three countries. All the information refers to a specific branch of training (education). In the three countries surveyed we find a diverse labour market background.

Since long ago, vocationalised (or pre-vocational) training in Africa has been criticised by numerous educational researchers. However, in most cases negative evaluations are based on a narrow economic viewpoint. Training has often been measured for individuals or social groups by cost-benefit analysis – the benefits calculated as monetary gains at the labour market, based on to what extent training has contri-buted to employment status.

Economic surveys of the kind referred to here seldom take account of the general societal situation. No concern is afforded to more general socio-economic trends prevailing in a particular country. Few reseachers seem to value the need for a more self-reliant development, or a more even spread of the material benefits of 'development' to disadvantaged groups.

While one hears outright denunciation of inefficient technical schooling/ training in Africa, it is admitted that there is a definite lack of empirical data to back up the view. The studies referred to here are part of an attempt to gather some more information on this line of school-training. It is essential for more studies of this kind to be initiated, for diverse education/training in various kinds of societies. One main objective would be to establish a monitoring follow-up system of former students/trainees from individual institutions.

From our studies we have established that some kind of technical diversification or vocational specialisation is not shunned by prospective candidates. Furthermore, schools, as well as training centres, seem to influence views and opinion among students/trainees. They seem to obtain a better understanding of technical matters, and an aspiration to explore this further.

In our case studies we can, at any rate, find some labour market advantages to be gained from the relevant studies/training. Naturally it is of various significance depending on the technical component offered, as part of the total course structure. Specialised vocational training and a diversified secondary education cannot be measured in the same way, based on labour market advantages.

One conclusion to be drawn from these studies is that there is probably a need for a somewhat varied set-up of technical diversification and more vocational training.

Finally, this paper has mostly discussed the purely educational/training aspects. However, often the main predicament is industrial development itself. The labour market does not develop at the same pace as educational output. At the same time, industrial production is at a comparatively low level for many African countries. Provision of manpower can only be one component of general economic development.

NOTES

1. For the section on Kenya reference can be made to Jon Lauglo and Kevin Lillis, *Vocationalizing Education – An International Perspective* (Oxford: Pergamon, 1988); J. Lauglo and A. Närman, 'Diversified Secondary Education in Kenya: The Status of Practical Subjects and Their Uses after School', *International Journal of Educational Development*, Vol. 7, No. 1 (1987), 227–42; Anders Närman, *Practical Subjects in Kenyan Academic Secondary Schools: Tracer Study* (Stockholm: SIDA Education Division Documents, No. 21, 1985); *Idem*, 'Technical Secondary Education and the Labour Market in Kenya', *Comparative Education*, Vol. 24, No. 1 (1988), 19–35; *Idem*, *Practical Subjects in Kenyan Academic Secondary Schools: Tracer Studies II Industrial Education (Three Years Follow-up)* (Stockholm: SIDA Education Division Documents, No. 39, 1988).
2. For the data on Tanzania see Anders Närman, *A Tracer Study Evaluation of the Moshi Vocational Training Centre (MVTC), Tanzania, Volume I* (Gothenburg: Department of Human and Economic Geography, Occasional Paper No. 6, 1987); *Idem*, *A Tracer Study Evaluation of the Moshi Vocational Training Centre (MVTC), Tanzania, Volume II* (Gothenburg: Department of Human and Economic Geography, No. 8, 1988); *Idem*, *A Tracer Study Evaluation of the Moshi National Vocational Training Centre (MVTC), Tanzania, Volume III* (Gothenburg: Department of Human and Economic Geography, Occasional Paper, No. 6, 1989).
3. For the brigade data see Ann-Catrin Emanuelsson, Eva Franzen and Anders Närman, *A Tracer Study Evaluation of the Botswana Brigades March–April 1988* (Gothenburg: Depart-ment of Human and Economic Geography, Occasional Paper, No. 11, 1988); Anders Närman, 'The Botswana Brigades; Some Preliminary Notes', *Botswana-Education, Culture and Politics* (Edinburgh: Centre of African Studies (1990), pp.159– 74.

Education of Farmers for Technology Transfer in the Third World

F. D. O'REILLY

Although there is a corpus of knowledge and technique called 'peasant science' (unjustifiably neglected in the literature) and peasant farmers do innovate and invent, innovation in traditional agriculture is usually analysed in terms of farmer response to innovations generated by central government, research stations or international agencies. Methods by which the centre communicates with the periphery, such as extension agency, demonstration plot, distance learning and the role of education of farmers in agricultural development are assessed. Such studies make it possible to identify the pattern and personnel of successful technology adoption and abandonment. The argument is illustrated by examples from Northern Nigeria, Thailand and Libya.

Technology Transfer

The Study of Innovations

Rogers[1] and Hägerstrand[2] can be regarded as the twin progenitors of innovation diffusion studies in agriculture, the former primarily concerned with the socio-economic, the latter with the spatial aspects. Rogers contended that socio-economic variables such as age, income, farm size, education level affect a given farmer's ability to innovate. In 1962 he postulated laws of innovation, but later, after travelling in the Third World, he changed these laws to 'generalisations'.[3] From this it follows that there will be significant differences in socio-economic characteristics between adopters and non-adopters, between earlier and later adopters, between discontinuers and non-discontinuers.

The study of innovation diffusion is of more than academic interest. Constraints of capital and trained technical manpower may make it impossible to service equally all farmers in a project area simultaneously. Thus a target group is supposedly carefully selected so that members of this group may serve as general change agents within their communities.

Dr F. D. O'Reilly is with the STD Forum, and the Education Authority of the London Borough of Newham.

129

Even if no target group is identified and selected, the process of innovation, with or without government intervention, will affect different farmers in different ways, so that after some years marked inequalities in the use of innovations in the farming community may be observed. Thus innovation studies may prove a useful predictive as well as descriptive and analytical tool.

Identifying the Peasant Innovator

Unfortunately, few of Rogers' aforementioned generalisations are conclusive (or even convincing), as his extensive bibliography demonstrates. The present writer made a specific study of age, selecting it as the simplest, least culture-specific variable and testing it as a proxy for other variables.[4] Early Rogers had thought that innovators were younger than non-innovators, a finding given prior backing by Gross[5] and given a theoretical justification by Davis.[6] Even fairly recently, Brown,[7] using secondary Mexican and Kenyan data, postulated that within geographical areas termed supply sheds 'factors related to age . . . are the major determinants of adoption'. Later Rogers was to state somewhat weakly: earlier adopters are no different from later adopters in age.

The present writer feels that part of the confusion arises from studying age as a solitary variable rather than as one linked to other socio-economic variables. If it has been demonstrated that early adopters in a particular society are younger, then this is the start rather than the end of the enquiry. The question to be asked next is what it is about being young which leads a farmer to innovate. In the USA, young farmers are believed to be dynamic, educated, thrusting, modern. These are decidedly characteristics that most south-east Asian farmers do not have: they are reticent, cautious, respectful, not presuming to air their wisdom in front of their elders; indeed in some cases the apparent farmer may not be the actual decision-maker.

It is possibly more fruitful to relate age to family and land holding size in a family cycle model of innovation adoption.[8] The model consists of three stages. Stage A is the young, recently wed farmer with a large number of dependents, a low potential labour force, an unfavourable dependent/worker ratio, a small land resource base and a high population density – the result is low innovativeness. By Stage B the middle-aged farmer with his mature and adolescent children (and possibly new wives) has a large potential labour force. This is, arguably, the most prosperous stage in the farmer's life and he may innovate with impunity. Stage C represents the phenomenon of the old farmer whose children have all married and set up their own households. The farmland may have been divided among them, leaving the old man with a rump holding, or the old man may remain the titular and increasingly ineffective head of a joint farm. At this stage material wants are few and the farmer's mind is generally on matters other – and higher – than agricultural innovations. Support for this model comes from Adewoye[9] in Kwara State of Nigeria, Apeldoorn, citing work in Northern Nigeria and Southern Niger,[10] and, most substantially, Barnett, studying the Gezira scheme in Sudan[11] who speaks of 'a tripartite scheme of labour situations' and states 'there is, therefore, an interesting and complex relationship between the tenant's biological life cycle and his position in regard to the organization of production'.

The implications are significant. Thus one is not looking for a progressive individual but an individual who at a certain stage of his life because of certain characteristics, partly engineered, partly fortuitous, has the ability to innovate, but these innovations (and resulting luxury purchases or items of conspicuous consumption) do not effect a total transformation, but may be discarded, abandoned or sold, when the individual's stage in the life cycle make their maintenance no longer feasible. Similarly, there is no creation of a new class of elite farmers or yeomanry with the ability to pass on intact and undivided their wealth to their descendants. Accordingly, it would also seem that the process of innovation diffusion must be repeated each generation among the *nouveaux pauvres*.

Evaluating Technology: Tractors in Peasant Farming

It is as important to evaluate the innovation as the mode of communication of the innovation, the process of innovation adoption or the characteristics of the innovator. In this context farm mechanisation is a particularly complex and emotive issue, involving economic vested interests as well as political and even spiritual overtones.

It is axiomatic that mechanisation improves markedly the yield per unit of labour, but it does not necessarily increase the yield per unit of land. Similarly it is not obvious that tractor mechanisation has much effect upon yield per unit of land as compared with animal draught power. Although mechanisation increased the productivity of the land by making possible the cultivation of land that would not otherwise be cultivable, it may also make some other areas of land 'uncultivable,' areas which are not accessible to machines. The margin of cultivation is a technological limit rather than a physical one. Mechanisation may bring down the margin of cultivation.

The relative efficiency and comparative productivity are also moot points. Clearly a machine does not clean a field entirely; not insignificant gleanings remain to the post-harvest harvesters of the Third World.

There is debate about whether tractors are more cost-effective than animals. Certainly, animals can be made more efficient by improving the equipment they power. It would seem reasonable to assume that the working life of a tractor is greater than that of a beast. However, machines in the Third World, aggravated by the problem of rust in humid tropical climates, are likely to become obsolescent more quickly than similar machines in the West. Conversely, such obsolete machines would be used much longer and there would be greater reluctance and ability to replace them. Breakdown is a serious problem for tractors in developing countries and tales abound of farmers 'nursing' rather than using their machines, of 'cannibalisation' and of spare parts being flown hundreds of miles.

Different types of land on a given farm may require different types of tractors. Similarly, it may not be feasible or sensible to mechanise one aspect of farming and then create a bottleneck by leaving other tasks unmechanised. Thus a demand for more equipment will be established, involving the farmer in ever increasing cost. A free economy results in a proliferation of brands, which means a need for more spare parts and trained mechanics and competition between agents in a field of relatively undiscerning customers.

If mechanisation extends the area of cultivable land, it will lead to an increase in the demand for labour. However, when mechanisation increases the intensity of cultivation in a given area, the effects on labour may be partly positive and partly negative: positively, it increases the requirements for planting, weeding and processing; negatively, it decreases the demand for labour in tilling the soil.

Technology is a product of a particular culture at a particular time and it is naive and rash to think other cultures can go shopping for suitable technologies and incorporate them into their economic and social fabric, leaving that fabric neither torn nor strained. Adopting a machine means adopting a whole complex of ideas and ideals, a new *Weltanschauung*.

Education for Innovation

Communication of Innovations

The term 'diffusion of innovations' has of late been increasingly replaced by 'communication of innovations' in that, whereas the former suggests a natural spontaneous spreading out, the latter term implies volition and makes clear, however tenuous the link or blurred the medium, that in essence the innovation process involves one human being communicating with another.

Communication of an innovation involves communication of the thing itself (the *Ding-an-sich*), be it water pump, tractor or new seed, and communication of the idea of the thing. Clearly object and idea are not necessarily communicated simultaneously: there may be a considerable time lag; ideally 'idea' should precede 'object', but it is not unkown for 'object' to enter a community before the 'idea of the object.'

Agents of communication can be formal or informal, involving on the one hand extension agents, foreign experts, farmers' associations and co-operatives and on the other hand neighbours or neighbouring villages. Sometimes formal and informal can combine, as when government targets and utilises local respected persons or big men, such as chiefs, teachers, policemen or priests. Such agents may use concrete here-and-now modes of instruction, as with the demonstration plot, the experimental farm or 'the innovation package', or may use a wide variety of media or distance learning techniques, including radio, TV, films, plays, puppetry, balla-deers, literature.

Farmers generally seem to pay more attention to the advice and example of their peers than to external authoritative agencies. In dealing with external agencies the repeated demonstration is preferable to verbal instruction or media messages. Use of media, such as film, seems to be largely unproductive, because even if the farmer can make sense of the unfamiliar, disembodied flickering images, they are rarely able to realise that it is meant to have any real relevance to their daily lives.

Of course the ideal of one extension agent per village is generally impractical, but farmers' complaints about the indifference or incompe-tence of extension agents occur too frequently to be discounted. Also it is not unknown for the ideas and objects to be deliberately impeded or unevenhandedly spread by designated agents of change. Ultimately access to new ideas and new objects may be crucial in farmers' ability to innovate, and whether they do or do not accept technology transfer may depend less on their level of education or socio-economic characteristics than on whether those who have been entrusted with the dissemination of the innovation see fit to share them with all the farmers under their care.

General Education

It seems a truism that education is necessary for development in agricul-ture and that investment in such education is prudent. This assertion needs qualification. In the first place it is possible for an educational system to maintain the socio-economic status quo rather than induce change. Secondly, that education is necessary for economic development is far from self-evident; there is a school of thought that believes that economic development precedes educational devlopment rather than vice versa. Education may be subdivided into general, higher and vocational education, being aimed at schoolchildren, scholars and workers respectively.

General education usually imparts a superficial acquaintanceship with academic subjects, as well as inculcating the values of state and society. Whatever value literacy *per se* may have, it is often assumed that it is a prerequisite for agricultural progress. Research findings, however, are contradictory. In the writer's opinion the link between literacy and agricultural development is by no means proven. In Northern Nigeria the rich and successful illiterate farmer or merchant is not uncommon; nor is the individual who is incapacitated by education. In north-east Thailand the majority of farmers are literate, but functionally illiterate in that they do not use their ability in daily life beyond reading song-sheets bought in the market-place. Then again it is possible to have literacy without comprehension. Much government literature intended for the rural classes is probably read without being understood: at times formal as opposed to vulgar vocabulary may confuse; even if all the words are understood, the message may not be clear.

Literate youths in north-east Thailand were found to have an aversion to reading,[12] whilst it has been demonstrated that literacy may be lost, if unused. The most comprehensive overview of literacy from an anthropological viewpoint is provided by Goody[13]: it is clear that far from being a liberating influence literacy may be an imprisoning and obstructive influence.

Technology Transfer Projects

The Sokoto Valley: The Graveyard of Technology

The Sokoto Valley in north-west Nigeria, at present the scene of the Sokoto-Rima Basin integrated development, is also a literal graveyard of machines from earlier colonial technology transfer initiatives from 1932 to just before independence in 1960.

The colonists sought to increase both productivity and production of rice, the former through the introduction of high yielding cultivars, the latter through mechanisation. The case of the new cultivars is dealt with elsewhere.[14] Here attention is focused on mechanisation.

Although as early as 1932 the colonists had attempted to introduce new technology when they gave farmers the opportunity to hire a pair of bullocks, implements and cattle at one shilling per day, it was in 1950 that tractors were introduced to the Zauro polder. There was wishful thinking that the Sokoto Valley could become a West African Irawaddy Basin. Low yields per acre did not seem to be the only obstacle to attaining this goal; equally disturbing was the relatively small area of the flood plain under cultivation (50 per cent around Birnin Kebbi). Some of the untilled area had never apparently been cultivated owing to the presence of a thick mat of *burugu* grass over heavy clay; elsewhere there was infestation with wild rice or *bau (Oryza barthii),* which could be eradicated only by ploughing to a depth of 8 inches. In both cases tractors seemed to be the answer and, according to reports, the first tractors needed to be replaced by heavier equipment. Some progress was made. Farmers were charged at the rate of 35 shillings per acre for land ploughed, 10 shillings of which they paid after ploughing, the remainder after the rice harvest. There was also a certain amount of re-allocation of land, producing more compact farms, but leading to a markedly inequitable land holding size distribution.

However, by the 1952–53 season targets were not reached owing to mechanical breakdown. The heavy tractors were replaced by lighter tractors which would get on the flood plain more easily and be more manoeuvrable. In this year also night ploughing was terminated because of farmers' complaints (this intriguing comment goes unexplained) and a bad harvest led to general defaulting on payments for ploughing. Farmers in Argungu openly stated that they did not desire any more tractor ploughing. A liaison officer was appointed to communicate between farmers and officers. Unfortunately, however, in the same year tractor ploughing had to close down early because of flooding and fear of being trapped.

Between 1956 and 1957 acreages ploughed by tractors had further declined and an overhaul of the machinery produced the verdict that the machines were good for another seven years. In 1958 farmers were approached by officials about raising ploughing charges. They resisted until persuaded by native officials. The last year of colonial rule saw perhaps inevitable decline. An entry in the Agricultural Station report for that year succinctly summarises the situation and also gives a sense of the air of frustration and, possibly, hopelessness, which was characterising the colonial officers. It reads: 'The scheme is possibly reaching the end of a cul-de-sac, where operational costs, total possible acreages and the maximum ploughing charge are all at cross purpose.'[15]

There are few people, including the colonial officers on the ground, who would declare that the scheme was anything but a failure. Davis laid the blame for failure on a variety of factors, including the high initial costs, the lack of skilled operators, the small size of the farms and the lack of genuine increase of yields.[16]

The present writer would add that there seems to have been a genuine communication gap between the farming populace and the administration; their 'philosophies of farming' were diametrically opposed. Secondly, the desire for change had not come from the farmers themselves, but was inspired from above. Thirdly, the colonists did see development as partnership, albeit an unequal one: government would supply the technology and the inputs but the farming populace would have to pay for these services out of cash derived from increased yields. This concept of a partnership with government into which the farmer could willingly enter or opt out seems to have been alien to the Sokoto farmers' ideas of what a fitting relationship between ruler and ruled should be. Most significantly, there was a basic financial dilemma in that the price demanded from the farmer was not in the first instance commensurate with the cost of the operation. What success the scheme had initially was due to the fact that the package was offered to the farmers at bargain prices. When real pricing was introduced, the farmers rejected the scheme, probably not simply out of parsimony, but because the hitherto realised and henceforth anticipated rewards did not seem commensurate with the increased inputs.

Possibly all of these problems could have been overcome, if it could have been demonstrated that mechanisation really 'worked'. However, the results of a survey carried out in the 1951–52 season showed that significantly higher yields were obtained on sandy soils through hand hoeing than by any mechanised means. Also, although deep ploughing by tractor attained the highest yields on clay soils, these high yields were not astoundingly higher than hand-hoe yields on the clay and were well below hand-hoe yields on the sandy soils. The traditional farmers had sagaciously selected the type of soil most suited to their technology and avoided the more intractable soils, as a sign of economic rationality rather than indolence. With existing markets for African rice (*Oryza glaberrima*) there was no pressing need for the farmers to extend their acreages onto the clays. The colonists introduced tractors not to improve the traditional multi-cultivar complex cultivation of African rice, but to replace it with a more homogeneous cultivar complex of higher yielding Asian white rice for the urban market. As in so many projects since, differing aims and priorities of managers and the managed doomed the technology transfer to failure almost from the start.

Northern Libya Resettlement Project

This study, carried out by Benzabih,[17] deals with a Libyan government settlement project in the Jabel al-Akhdar region of north-east Libya, an area of predominantly rain-fed agriculture and of former colonial (Italian) agricultural settlement. Out of this background arise two major problems: the raising of productivity in edaphically marginal areas and the settlement of traditionally nomadic pastural people into a rural milieu of alien genesis.

One of the government's main objectives was to create a stable agricultural community; to this end it entered into benevolent agreements towards the repayment of mortgages. The project authority adopted the mixed farming system, with the small family farm as the cornerstone in agricultural development policy. Based on land classes various farm grades were established. Grade I farms constitute the most favoured edaphically and climatically and thus achieve the highest cereal yields; Grade III farms are the most marginal in the project area. When the average income of different farm groups is considered, however, considerable and surprising discrepancies are evident. The startling contradiction emerges that the farms with the poorest physical resource base are in fact the most profitable because of their rough grazing. In fact farmers in Grade I fell below the national average GNP per capita (for 1981), whereas the average minimum income for farmers in Grade II and Grade III is above this line. Moreover, since labour inputs are one of the principal inputs, the grotesque anomaly arises that the farm income is inversely related to labour inputs, that increased effort gives diminished reward. It is clear that income from livestock enterprises in the area is the single most important source of farm income in the area and that the farm settlement model has proven unviable.

Although the project-approved farming system had been widely demonstrated to the farmers of the region, farmers invariably chose to invest their time and effort in farming activities they preferred. A variety of reasons may be forwarded for this lack of success: the cessation of government grants; the inefficiency of institutional organisation; the low level of interaction between farmer and institution; under-used credit facilities; impotent co-operation; agro-industries not integrated with local production; counterproductive marketing monopolies; ill-equipped extension services; high rate of illiteracy. Because of this institutional weakness, it may be argued, farmers must conduct the running of the farms on the basis of their own past experience. However, farmers' relative neglect of approved farming methods is as much for rational, positive reasons as simply a reaction to poor organisation: farmers' lack of knowledge of certain 'delicate' crops is coupled with their sure knowledge of the higher net returns of sheep rearing than other activities.

Revaluation of Values

It has been pointed out in this paper that schemes often founder through conflict, overt or otherwise, between managers and managed. Hence any revaluation of values must aim at revaluating these two categories.

Before scrutinising the peasant, those concerned with the development field should perhaps individually and collectively engage in introspection. The questions 'why are we here?' and 'why are we promoting this line of development?' may, if tackled honestly and in depth, provide uncomfortable answers. In this context it should never be forgotten that there is such a thing as a 'development industry' and development can provide a career equally lucrative and satisfying as any of the older established professions, though often without their stringent entrance requirements on codes of conduct.

Turning to the peasant, one has to accept from the outset that one is dealing with a rational human being. His actions may not indeed be always economically or environmentally sound or indeed always in his best interest but they are almost always reasoned and considered. His prime motivation is survival – all the rest is secondary and, arguably, in most cases he has more knowledge of what is conducive to his own interests than the instant expert or even the long-term mud-on-the-boots expert, who, numerous writings and heartfelt dedication notwithstanding, will not suffer, if his developmental directives prove mistaken.

Between manager and managed, between expert and farmer there is inevitably a sizeable army of intermediate level officials – enumerators, extension officers, data analysts, interpreters. These, frequently ill-trained and underpaid, are not paid much attention to in the literature but, in the writer's opinion, the quality of such staff may sometimes be the single most important factor in the success or failure of projects, and certainly the quality of collected data makes even the most rigorously electronically analysed data suspect. More pay and training might improve their performance, but, arguably, the crux of the matter is that they do not really know why they are doing the tasks assigned to them and fail to relate their own small role to the large development project.

The element of time is crucial in monitoring and evaluation. Analysts may come up with markedly disparate conclusions depending upon when in the life of a project they make their observations. Early observations may reveal teething trouble or subsidised success; later observations a comprehensive transformation or obsolescence or breakdown. Presumably all technology transfers are intended to continue and be self-sustaining, when all the initial aid, training, expert personnel and grants have been removed. This makes it doubly necessary at the outset that there is a uniformity of purpose or at least mutual respect and communication between manager, managed and their intermediaries.

NOTES

1. E. M. Rogers, *Diffusion of Innovations* (New York: Free Press of Glencoe, 1962).
2. T. Hägerstrand, *The Propagation of Innovation Waves* (Sweden: Lunds Universitet, 1952).
3. E. M. Rogers and F. F. Shoemaker, *Communication of Innovations: A Cross-Cultural Approach* (New York: Free Press of Glencoe, 1971).
4. F. D. O'Reilly, 'Age and the Peasant Innovator' (London: SOAS Occasional Paper No. 5 (New Series), 1983).
5. N. C. Gross, 'The Diffusion of Culture Traits in Two Iowa Townships', unpublished M.Sc. thesis, Iowa State College, 1942.
6. A. Davis, 'Technicways in American Civilization', *Social Forces*, 18 (1940), 317–30.
7. L. A. Brown, *Innovation Diffusion: A New Perspective* (London: Methuen, 1981).
8. F. D. O'Reilly, 'Rural Children's Education and the Green Revolution in Nigeria', *Journal of General Studies*, Vol. 3, No. 1 (1982), 193–9.
9. A. Adewoye, 'The Process of Innovation in Peasant Agriculture in Arandun, Kwara State' (Kano, Nigeria: BUK undergraduate dissertation, unpublished, 1980).
10. J. van Apeldoorn, *Perspectives on Drought and Famine in Nigeria* (London: George Allen & Unwin, 1981).
11. T. Barnett, *The Gezira Scheme: An Illusion of Development* (London: Frank Cass, 1977).
12. J. W. Pilgrim, 'Social Planning for Rural Development', draft, Bangkok, 1972.
13. J. Goody (ed.) *Literacy in Traditional Societies* (Cambridge: Cambridge University Press, 1981).
14. F. D. O'Reilly, 'The Sokoto Valley Rice Project: 1926–1960', *Kano Studies*, NS 2 (4) (1986), 167–77.
15. *Northern Nigeria Agricultural Station Reports*, Vol. 1, Part 1, Sokoto, various years.
16. L. L. Davis, *Status of Rice Production Nigeria*, USAID, consultant report, 1962.
17. H. A. Benzabih, 'Environmental and Socio-Economic Constraints on Rain-fed Agricul-tural Land Settlement Projects in Marginal Climatic Zones: A Case Study of the Jabel al-Akhdar Upland, Libya', unpublished Ph.D. thesis, SOAS, London, 1987.

Transfer of Applied Technology

GERNOT WAPLER

The focus of this study is the training of Third World personnel in Italy, with the German experience being presented as a comparative foil. It maintains that there is a great opportunity for Third World students to be two-way intermediaries (between developed and less developed nations) and change agents within their own countries. However, the opportunities are often lost more as a result of lack of infrastructure, back-up, sensitivity and appropriateness in the donor country than failings of the students. Despite its shortcomings Italy is seen as a 'more appropriate' model: owing to historical and geographical factors, it can provide appropriate economic and technical models for transfer; at the same time it provides an easier economy into which foreign students can integrate, often informally, with results that are not entirely beneficial; finally, it provides a magnet for a significant Third World brain drain. Italy may thus be resuming its traditional transitional position between North and South and West and East, though Middle Eastern students are in decline, whilst African students are on the increase. In spite of the arguably lesser appropriateness as a transfer-model of the German situation, that country's more development-oriented programmes furnish Italy with a fitting exemplar.

The transfer of applied technology and knowledge is a viable strategy in integrated development programmes, whilst the epithet 'applied' has become a *sine qua non* for development. It is necessary to design development programmes that fit in with the social, technical, educational and environmental backgrounds in the developing countries. In brief, no

Gernot Wapler is with the Berlin Institute of Environment Research, Berlin, Germany.

The author would like to thank Dr F. D. O'Reilly of STD Forum (Glasgow) and Plashet School (London) for reading this paper on his behalf at the conference and also for valuable assistance in producing the English version of this paper.

The views expressed in this paper are entirely the author's own rather than 'official' German or Italian views. Similarly, they are not based on a thoroughly comprehensive overview or detailed survey of the situations in Germany and Italy. They should be viewed as suggestive rather than conclusive, as opening up a major and important area of debate and pointing the way to further research and more prudent policy-making.

project should be designed that imposes alien concepts on the populations of these countries. This implies the need both for a constant process
of reviewing and for the identification of new means and new opportunities for technical co-operation. The field of technology and knowledge
transfer through students from developing countries studying at European
universities provides an example of lost opportunities.[1] A comparative
analysis of the situation in Italy and Germany is of special insterest,
because in both cases the number of foreign students is quite high and
their languages are not spoken in the home countries of the students. A
new and unfamiliar language requires a high motivation on the part of the
student, although clearly not all European languages pose equal difficulties. Arguably, Italian lends itself to at least a superficial rapid
mastering. Italy has the further advantage in that there are no restrictions
or 'quotas' for the admission of foreign students. Traditionally, most of
Italy's students have come from the Middle East, because these nations
have had greater financial resources to sponsor study abroad. In the last
few years the political crises in these countries has provided an additional
strong reason for studying in Italy, a country which has been known for
its tolerance towards people with other cultural backgrounds.[2] Nevertheless, the total number of Middle Eastern (ME) students has actually
decreased of late. A further Italian advantage is that it offers more job
opportunities in the informal sector, which enables students to finance
their study, should the 'usual cheque' from home not arrive. It is African
students who mostly avail themselves of such opportunities. The German
case is of interest because of the worldwide reputation of Germany's
strong economy and the high level of its technology. Germany also offers
many scholarships, which are powerful magnets for Third World students.
In this paper emphasis is placed on the Italian example, with the German
case providing a foil.

Students from the Third World have become more and more conscious
that studying abroad is not only a way of improving one's personal
economic situation, but that it is also a means of contributing to the
development of the home country. This makes such students suitable
partners for international co-operation. Students studying in Europe,
who are committed to the development of their home countries, constitute ideal partners for the integrated transfer of applied knowledge, as
they are more cognisant of the perceptions and mentalities and are more
sensitive to the desire for increasing autonomy in their places of origin
than Europeans can ever be. At the same time such students have gained
familiarity with Western technical and scientific methodologies and thus
are able to play an efficient role in research programmes and technological developments which serve their countries. Identification of the

structures of autonomy in economic perception and behaviour is a prerequisite for the successful transfer of technology and knowledge. This
can be achieved only by means of reciprocity, that is, bilateral co-
operation and interdisciplinary research [*Wapler, 1988*]. In this context
the West depends on the co-operation of students from developing
countries, who are in a position to exercise a critical sifting of technological development methods. Thus the Western experts are not the
only ones who state whether a technology is appropriate. The potential
contribution of these students depends, however, on the social and
economic situation each individual student faces during his studies. There
are significant differences between Italy and Germany in this respect.

A research group in the Italian region of Umbria studied the problems
of integration of foreign students in an Italian university town [*Brunelli et
al., 1989*]. It was stressed that the students become 'individuals between
two cultures', a condition aggravated by language problems and that this
phenomenon may have an effect on the success of their studies. Several
social associations operate in the fields of cultural exchange and offer the
foreign students relief from stressful situations. Very often the economic
and professional situation of the students deteriorates. On the basis of the
findings of the research group and of the writer's own experience it can be
stated that the students enter into a sort of cycle of conflict situations:

(1)   In the beginning the student sees his study in a developed country as
       a chance to enter into the mechanism of personal economic and
       cultural improvement.
(2)   The student adopts 'a strategy of adaptation' even if it means
       endurance in the face of the new cultural system and new techno-
       logical models. This may even lead to the formation of a new
       personal *Weltanschauung*, linked to a certain degree with the fact
       that he has to express himself in another language.
(3)   This leads to the phenomenon of social disorientation and lone-
       liness and has a negative impact on the personal capability to
       optimise study and career. At this stage, frequently, the student will
       change university or even abandon his studies, facilitated by the fact
       that in Italy a lot of part-time or low-income job opportunities are
       offered in the informal sector of *economia sommersa* (submerged
       economy), including seasonal or casual labouring on Italy's rather
       labour-intensive farms.
(4)   After a phase of uncertainty the student is faced with the decision
       whether to settle in the host country, thereby contributing to the
       brain drain. Alternatively he may develop a new consciousness of
       the needs and assets of his home country. (Such decisions are made

TABLE 1
STUDENTS FROM DEVELOPING COUNTRIES AT ITALIAN UNIVERSITIES AND
OTHER INSTITUTIONS, 1982/83 TO 1987/88

	University	Other Training Institutions	Total	Post University Courses
1982/83	10,642	8,402	19,044	n.a.
1983/84	11,581	7,607	19,188	n.a.
1984/85	11,612	7,502	19,114	n.a.
1985/86	12,020	6,872	18,892	685
1986/87	10,500	6,787	17,378	598
1987/88	9,634	6,550	16,184	482

Source: Amicizia, Studenti Esteri, Anno XX–XXV, Rome.

more difficult when, as happens not infrequently, Third World students have entered into marital or at least emotional relations with citizens of the host country – arguably, the response differs depending upon whether it is the male or female partner that is of Third World origin.)

This last stage is the moment when the question of knowledge transfer comes in with renewed vigour. The student may realise new objectives in his study and research and will be able to channel his efforts and interests. Italian institutions of technical co-operation have become more and more aware that these potential 'conveyors of development' can be effectively utilised for the implementation of development programmes at international, bilateral, national and local levels. It is necessary though that the acquisition of knowledge is accompanied or followed up by advanced and post-graduate courses with special reference to developing countries. The belated awareness of this fact partly accounts for the significant brain drain from the Third World to Italy. It should also be noted that in Italy the structure of study at university is designed to offer a broad and general acquisition of knowledge rather than aiming at early specialisation and practice-orientation. Foreign students who had intended to return to their own countries did not therefore find much support in the preparation for their future tasks – at least until the beginning of the last decade. Universities like Turin, Trieste, Perugia and Milan have now made a special effort to review their programmes towards a more development-oriented teaching.

TABLE 2

ITALY: UNIVERSITY STUDENTS FROM DEVELOPING COUNTRIES BY AREA OF
ORIGIN, 1982/83 TO 1987/88

	Middle East	Other Asia	Africa	Latin America	Total
1982/83	7,620	303	1,884	835	10,642
1983/84	8,140	287	2,200	954	11,581
1984/85	7,992	310	2,349	961	11,612
1985/86	8,222	336	2,539	923	12,020
1986/87	7,078	307	2,431	684	10,500
1987/88	6,338	287	2,304	705	9,634

Source: Amicizia, Studenti Esteri, Anno XX–XXV, Rome.

Statistics show a slight decrease in the total number of Third World students from 1982/83 to 1987/88 oscillating around 10,000. A further 6,000 to 8,000 were enrolled in secondary and technical pre-university training institutions, whilst around 500 to 600 were enrolled at post-university training courses (see Tables 1 and 2). The drop in the total figures is mainly due to a significant decrease in the flow from Iran. Among the African areas of origin North Africa, Nigeria and the former Italian colonies of Somalia and Eritrea represent about 70 per cent of the total.

Table 3 highlights the extent to which such studies meet the urgent demands of applied knowledge and technology transfer. In fact subjects like medicine and architecture – seen as prestigious and lucrative branches of study – still dominate over the development-oriented subjects such as agriculture, natural science and engineering. This applies mostly to ME-countries. Whilst the number of ME-students is decreasing, the number of African students is steadily increasing, not only in medicine, but also in agriculture, economics and natural science. This fact raises hope that studies are now contributing towards the overall effect of knowledge transfer. In the last few years the importance of advanced courses for experts and professional staff from developing countries has been realised by the educational planners, while simultaneously there has been a dramatic drop in the number of students enrolled in post-graduate courses.[3]

TABLE 3
ITALY: ENROLMENT OF UNIVERSITY STUDENTS FROM DEVELOPING
COUNTRIES BY FACULTY AND PLACE OF ORIGIN, 1982/83 AND 1987/88

	Middle East		Other Asia	Africa	Latin America	Total	
	Abs	%				%	Abs
Total 82/83	7,620	100	303	1,884	835	100	10,642
Medicine, Vet. & Pharmac.	4,009	53	155	578	265	47	5,007
Architecture	1,495	20	25	232	67	17	1,819
Phil., Political Sc., Law	424	5	29	182	220	8	855
Economics & related Sc.	156	2	10	167	65	4	398
Engineering	1,007	13	50	489	128	16	1,674
Nat. Science	293	4	28	106	59	5	486
Agriculture	220	3	6	117	27	3	370
Others	16	—	—	13	4	—	33
Total 87/88	6,338	100	287	2,175	705	100	9,634
Medicine, Vet. & Pharmac.	3,537	56	125	761	184	48	4,607
Architecture	1,320	21	25	333	94	18	1,772
Phil., Political Sc., Law	376	6	60	216	185	8	837
Economics & related Sc.	80	1	10	250	52	4	392
Engineering	642	10	30	340	81	11	1,093
Nat. Science	231	4	10	182	37	5	460
Agriculture	108	2	8	167	17	4	300
Others	44	1	19	55	55	2	173

Source: Amicizia, Studenti Esteri, Anno XX–XXV, Rome.

TABLE 4

ITALY: ENROLMENT OF STUDENTS FROM DEVELOPING COUNTRIES IN
POSTGRADUATE COURSES BY AREA OF ORIGIN, 1984/85 TO 1987/88

	1984/85	1985/86	1986/87	1987/88
Total	2,528	685	598	482
Asia	694	152	113	60
Africa	1,113	265	328	224
Latin America	721	268	157	198

Source: Amicizia, Studenti Esteri, Anno XX–XXV, Rome.

Advanced training efforts focus on professional courses, mostly sponsored by the Italian government, but also by industrial associations. The
number of participants on these courses has exceeded the number of
postgraduate students at university for several years. This demonstrates
that the official strategies are putting emphasis on technology transfer
rather than on knowledge transfer, thus not fully utilising the potential
offered by the high number of Third World students enrolled at universities. From the fact that many courses are held in Italian, it can be
deduced, however, that they are especially offered as a sort of follow-up
training, to former students at Italian universities, who have in the meantime gained some professional practice in their home countries.

Several courses do aim at an optimal interaction between transfer of
knowledge and transfer of technology based on co-operation with either
international academic and research institutions or industrial establishments. This is facilitated by the development of a strong medium-scale
agricultural and machinery industry in Italy offering a large range of
innovative production procedures that can conveniently be taken over by
developing countries because of their relatively small scale and because
of comparable environmental conditions. Thus Italy's experience with
aridity, irrigation technology, marsh drainage and rice cultivation all
have an immediacy and a relevance that most Western European countries cannot emulate. Italy may in fact be resuming its traditional historical role of bridge between North and South and West and East. A good
example of this relevance is provided by the Water Resources Research
and Documentation Centre (WARREDOC) near Perugia, which was
designed to offer an alternative to the standard development projects
with the input of costly machinery and hardware and the employment of

TABLE 5

ITALY: ADVANCED COURSES FOR DEVELOPING COUNTRIES BY NUMBER OF
PARTICIPANTS, SUBJECT AREA, WORKING LANGUAGE AND DURATION, 1988

Subject Area	Number of Courses	English, French Spanish		Italian		Total Participants
		< 6 months	6–12 months	< 6 months	6–12 months	
All	73	706	322	65	805	1,898
Agriculture	22	160	45	20	357	582
Banking	4	85	55	—	—	140
Economics	4	20	95	—	20	135
Energy	15	230	67	15	25	337
Teaching	6	126	—	—	20	146
Industry, Cr.	13	—	60	—	239	299
Public S.	3	50	—	—	44	94
Others	6	35	—	30	100	165

Source: Ministero Affari Esteri [1988], Cooperazione Italiana. Repertorio Corsi di Formazione, Rome.

highly paid experts from the donor country [WARREDOC, 1990]. The
Centre was instituted in 1984 in order to promote research activities and
the exchange of ideas in the fields of water resources, medicine, food
production, environmental protection and computer-oriented management. Case studies are being carried out in several parts of the world, for
instance in Borno State and the Port Harcourt area in Nigeria in nine
subject areas – namely, mini-hydro pre-assessment and design, hydrological problems of channels and sewage systems, general hydrological
problems, reservoir management problems, hydro-geological issues,
agro-meteorological issues, water supply, water quality, planning and
management issues. The training exemplifies the 'sandwich system', as
the participants, professionals in both governmental and private institutions and agencies, are carrying out the research in close contact with
their home countries. Thus it can be regarded as an effort to overcome the
'one-way' direction of North-South co-operation.

The German model is based on an educational system that is more
practice-oriented and offers more follow-up facilities for the students
from the developing countries. On the other hand, the structure and

TABLE 6

WEST GERMANY: UNIVERSITY STUDENTS FROM DEVELOPING COUNTRIES
BY AREA OF ORIGIN AND SUBJECT AREA, 1985/86

Subject	Asia*	Africa	Latin America	Total
Total	20,588	4,218	3,129	27,935
Cultural Science & Language	2,639	625	763	4,027
Sports	111	27	47	185
Law, Economics & Social Science	2,606	739	483	3,828
Natural Sciences	3,696	632	505	4,833
Medicine, Vet.	1,879	241	172	2,292
Agriculture	788	307	151	1,246
Engineering	8,029	1,560	773	10,362
Arts	800	80	219	1,099
Others	40	7	16	63

Source: CIM [1987], 'Studenten aus Entwicklungsländern an deutschen Hochschulen', Arbeitsmaterialen 7.
* Excluding Asiatic Turkey

conditions of the German economy, highly export-oriented, are completely different from those found in the home countries of the students. Table 6 shows that in the term 1985/86, 27,935 students from developing countries were enrolled at German universities; another 3,170 were enrolled at university colleges.

By comparison with the Italian situation, there is clear evidence that there is a more equal distribution in the subject areas. Medicine does not dominate; engineering takes first place, mainly chosen by Middle East students. Engineering also includes architecture. This contrasts with the Italian study of architecture, which puts more emphasis on the design aspects than the technical ones. The objectives of the students studying in Germany are obviously more oriented towards the acquisition of technological knowledge and thereby more development-oriented. This is further supported by numerous postgraduate courses at universities, especially designed for students from developing countries, which aim to prepare the students for the tasks in the home country. In 1988 2,500 scholarships were granted to postgraduate students (comparable to the number in Italy, but with longer duration) [*Deutscher Akademischer Austauschdienst, 1989*]. Government promotes several reintegration programmes for the return of the students by means of scholarships and follow-up activities. The German model is thus characterised as a highly integrated method of technical co-operation intended to utilise fully the potential contribution to Third World development of students from developing countries. Advanced courses in professional and management training exemplify this approach. This notwithstanding, problems do remain with the German model. It is, moreover, not easily transferable to other Western countries such as Italy, because of the different economic and social conditions. It does offer, however, a sort of frame of reference by which the Italian situation can be examined.

In conclusion, study in Italy offers for the student from a developing country several advantages as far as the chance of reintegration into the home society is concerned. The level of Italian technology may also be quite appropriate for application in many development-oriented fields. The students will certainly play a more active role in the development of their countries, if there is a concerted effort to integrate special seminars, workshops, courses and collaborative research programmes into the Italian university system. The fact that economic development is anticipated by technological development suggests that the acquisition of technological knowledge by the students will have to be improved in order to put them into a position to serve their home countries efficiently. Personal motivations have to be enriched by long-term development-oriented objectives. The following conditions are prerequisites for the realisation of these objectives:

(1) Financial and logistic assistance, aimed at improving the social and economic situation of the students and making the university system more accessible, transparent and relevant.
(2) Improvements in the direction of a more practice-oriented curriculum development and provision of more post-graduate specialised courses of applied technology.
(3) Development of a reintegration programme which improves the student's impact on his home country and helps to block the brain drain from these countries.

With these modifications the sojourn of Third World students in Italy would provide potential for a meaningful educational exchange with a view to building up a new generation of partners and hence more reciprocity in North–South co-operation.

NOTES

1. Training of Third World students in Western European academic and vocational institutes is of course only one method of technology and knowledge transfer. Others include Western aid (including technical personnel) in the setting up of training centres in Third World countries or of centres of excellence in certain nodal Third World countries serving larger regions, such as ECOWAS or ASEAN, or the financing of Third World students to study in other Western nations (perhaps with more appropriate or cost-effective courses or more familiar languages). United Kingdom aid is being increasingly directed towards the former method; Finnish aid to the latter. There is a wide-ranging and controversial debate about the relative efficiency and cost-effectiveness of these various methods.
2. Doubtless students of different and culture ethnicity experience varying degrees of accep-tance or antagonism from one European country to another. Third World students appear to feel 'more at home' and experience less deep-seated hostility or racism in Italy than in some other Western countries. At the same time certain sectors of the Italian economy, such as farming and rented accommodation have become, in certain parts of the country, far more dependent on Third World students' labour and/or income than is arguably the case in other Western countries.
3. The terms 'he' and 'his' are used throughout to refer to Third World students for convenience and to avoid syntactical ungainliness. These terms should be taken to embrace 'she' and 'her'. There is no intention of suggesting that only male Third World students can be effective 'conveyors of development'. At the same time it must be accepted that the experiences and aspirations of male and female students may differ; so may the precise ways in which they can most effectively contribute to technology and knowledge transfer from host to home country. Such issues of gender, however, though important, are outside the brief of this study.
4. The statistics shown in Tables 1 to 4 are the result of special surveys carried out by UCSEI (Central Office of Foreign Students in Italy). Whilst the statistics on University enrolment include all Italian universities, the statistics on post-graduate courses, secondary and technical institutions and vocational centres correspond to 80–100 per cent of the total.

REFERENCES

Amicizia, Studenti Esteri, Anno XX–XXV, Rome.

Brunelli, L., Bussini, O., Cecchini, C. and L. Tittarelli, 1989, *La presenza Straniera in Italia. Il caso dell'Umbria,* Milan.

CIM, 1987, 'Studenten aus Entwicklungsländern an deutschen Hochschulen', *Arbeitsmaterialen* 7.

Deutscher Akademischer Austauschdienst, 1989, *Scholarships for Post-graduate Courses,* Bonn.

Ministero Affari Esteri, 1988, *Cooperazione Italiana. Reportorio Corsi di Formazione,* Rome.

Wapler, G., 1988, 'Considerazioni sull'Impostazione Teorica e Metodologica della Geografia Applicata ai Problemi del Terzo Mondo', *Rivista Geografica Italiana,* Anno VC, 4.

WARREDOC, 1990, 'International Advanced Course on Water Resources Management: Water for Health, Water for Food, Water for Energy', Villa la Colombella, Perugia.

Manpower Constraints in the Industrial Recovery of Post-War Iran

HASHEM ORAEE AND ALI HAERIAN

Since the end of hostilities between Iran and Iraq a considerable amount of attention has been given to the question of manpower development in order to facilitate the reconstruction of Iranian industry. With the proposed increase in industrial output in the present five-year plan the productive capacity of many existing plants will be increased, unfinished projects completed and new ones implemented.

A major obstacle facing the authorities is the development of manpower with the necessary knowledge and skills to cope with the proposed enormous expansion in the industrial sector and also to ensure a consistent rise in efficiency and productivity.

This paper examines the manpower requirements in the light of the present five-year plan. The increasingly important role of education and training at universities and other higher education institutions is highlighted. The current state of research and development in the country is also presented. The paper concludes by examining the planned increase in research activities at various levels during the next five years.

Introduction

A few months after the election of the present government in Iran, the details of a five-year plan covering the period 1989–1994 were put together, following intense debates at various levels, and its main proposals were approved, with minor modifications, by the Iranian Parliament. The major aim of the plan is to get the industrial sector of the economy back on the move, now that the hostilities have ceased. For this reason it has attracted a considerable amount of political as well as psychological importance and the government appears to be fully committed to seeing it through successfully.

The authors, Hashem Oraee and Ali Haerian, are in the Faculty of Engineering, Ferdowsi University, PO Box 91775–1111, Mashad, Iran.

151

Over the past decade, a significant proportion of the industrial and agricultural sectors of the economy have been attracted to the service industry due to financial incentives, as a result of which the whole economic pattern has moved away from production to the provision of services. In addition to this, a staggering population growth rate currently standing at about 3.9 per cent has meant, in some cases, acute shortages of some basic commodities.

With the industrial recovery of the Iranian economy as its main objective, the proposed programme intends to increase the production level of both the industrial and agricultural sectors, and this involves the use of advanced technologies. The construction of a number of dams for irrigation purposes as well as generating hydroelectric power, power stations, and other capital intensive projects is planned in order to facilitate an increase in the production capacity of the economy.

In other words, the plan proposes to increase the production level of the industrial and agricultural sectors of the economy at the expense of the service sector, thereby easing, at least partly, the pressure on the demand for many products. At the same time tight monetary policies are to be carried out by keeping the supply of money under close control and aiming to achieve a balanced budget within the second half of the plan.

Manpower Requirements

One of the major obstacles for the authorities in Iran is the lack of technical skills at various levels in industry. Traditionally, industry in Iran has attracted turn-key projects involving the transfer of know-how from the developed world. In many cases, poor training and lack of supervision has led to improper use of machinery and a reduction in the projected level of output both in terms of quality and quantity.

Unlike many of the Gulf states in the region where the major manpower constraint is the small size of the local population (compared with the foreign workforce), the main problem in Iran is one of educating, training and, in some cases, retraining of the relatively unskilled and undisciplined workforce.

The lack of interest in new technologies developed in various parts of the world and financial restrictions imposed by the war has meant that, except in certain defence-related industries, major structural training programmes have to be carried out. At present, comprehensive training programmes at all levels are quite limited and the government plans to concentrate on education and training by expanding the role of universities and other higher education institutions.

TABLE 1

PERCENTAGE OF AGE GROUP ENROLLED IN EDUCATION

Country	Primary (6–11)		Secondary(12–19)		Tertiary(20–24)	
	1965	1985	1965	1985	1965	1985
Ethiopia	11	36	2	12	0	1
Bangladesh	49	60	13	18	1	5
China	89	124	24	39	0	2
Egypt	75	85	26	62	7	23
Brazil	108	104	16	35	2	11
Iran	63	112	18	46	2	5
Spain	115	104	38	91	6	27
U.K.	92	101	66	89	12	22
Japan	100	102	82	96	13	30
Canada	105	105	56	103	26	55

Source: *Research and Development Programs in the Five-Year Plan of the Islamic Republic of Iran*, Ministry of Culture and Higher Education, 1988.

Education and Training

Although there has been an impressive increase in the total capacity for university education in Iran over the past few years, the high rate of population growth coupled with the large size of the young population of the country has led to increasing competition in the annual university entrance examination which takes place at the national level.

A comparison of the percentage of various age groups engaged in education in different countries is shown in Table 1, from which it is evident that between 1965 and 1985 the proportion enrolled at the primary level (age group 6–11) almost doubled. The corresponding increases in secondary education (age group 12–19) and tertiary education (age group 20–24) are 3 and 2.5 fold respectively. It is also evident from Table 1 that in the tertiary education sector this proportion is far lower than in any of the developed or developing countries shown in the table.

Extensive efforts are being put into education at various levels throughout the country. This is clearly evident from Table 2, which shows

TABLE 2

COMPARATIVE EXPENDITURE ON DEFENCE, HEALTH AND EDUCATION
(1985)

Country	Population (x 1000)	GNP/Capita (US $)	Defence %of GNP	Health %of GNP	Education %of GNP
Ethiopia	42 271	110	9.3	1.4	3.0
Bangladesh	100 592	150	1.7	0.4	1.2
China	1 041.094	310	7.0	1.4	2.8
Egypt	47 108	710	8.5	1.2	4.3
Brazil	135 539	1 640	0.8	1.6	4.0
Iran	45 160	1 778	13.3	1.6	7.5
Spain	38 730	4 360	2.4	4.6	2.5
U.K.	56 539	8.390	5.4	5.4	5.1
Japan	120 579	11 330	1.0	4.6	5.1
Canada	25 414	13 670	2.3	6.4	7.4

Source: Research and Development Programs in the Five-Year Plan of the Islamic Republic of Iran, Ministry of Culture and Higher Education, 1988.

the percentage of GNP spent on defence, health and education in ten different countries, and shows that the expenditure on education as a percentage of GNP is highest in Iran, followed closely by Canada and then Japan and the UK.

At the end of the five-year plan, the total number of students graduating from universities each year, which currently stands at about 27,000, will be increased to 40,000. This is coupled with a planned increase in the total number of university students by 65,000. During the past decade, the tertiary education sector has suffered a severe shortage in the number of academic staff, mainly due to a considerable proportion leaving the country in the early 1980s. However, in the latter part of the decade the trend had reversed and there was a steady increase.

The total number of university students in the academic year 1986–87 was 170,614, of whom 71 per cent were males (see Table 3). Comparing the number of students in the academic year 1986–87 with that of 1985–86 shows an increase of 12.6 per cent in the number of students engaged in university education at various levels.

The number of graduates for the academic year 1985–86 was 28,971, of whom 32.3 per cent were females (see Table 4). The number of students graduated from state universities and higher education institutions in the academic year 1986–87 showed an impressive 45.3 per cent increase over the previous year.

It is worth noting that special attention has been paid to the training of highly specialised manpower at postgraduate levels. In the academic year 1977–78, the proportion of students studying at postgraduate level stood at 10.7 per cent. This proportion had increased to 16.4 per cent in the academic year 1986–87.

More emphasis is being placed on training in the universities. Students of engineering and science courses are particularly encouraged to attend training programmes of a few months duration at various state-owned or private industries. Training courses are also organised by the relevant Ministries on a regular basis and industries are offered financial and other incentives to arrange for their workforce to attend these programmes. It is proposed that promotions at both shopfloor and managerial levels be directly linked to successful completion of such courses.

Research and Development

The picture is different on the R&D front. Taking the traditional criteria for evaluating the state of research and development, namely, the percentage of GNP spent on research in a year and the number of researchers per one million of population, the figures are not encouraging at all. Figure 1 shows government expenditure as a percentage of GNP spent on R&D during the period 1979–88.

When this is compared to the amount of R&D expenditure in other countries, the disparity becomes more alarming (see Table 5). Moreover, from the distribution of R&D expenditure shown in Table 6, it is evident that a considerable proportion of the total R&D expenditure has been devoted to industrial research.

As for the number of researchers per million of population, while this number was 4,800 and 3,300 in 1985 for Japan and USA respectively, the average for the developing countries is about 500. In 1987 this number was a mere 82 in Iran (assuming that one-third of all university academic staff are engaged in R&D).

TABLE 3

NUMBER OF STUDENTS BY TYPE OF INSTITUTION,
LEVEL OF STUDY AND SEX, 1986–87

Type of Institution	No. of Institutions	Total			A.A.[1] Programmes		B.A. or B.S.[2] Programmes		M.A. or M.S.[3] Programmes		Professional[4] Doctorate		Ph.D[5] Programmes	
		Total	Female	Male	Total	Female	Total	Female	Total	Female	Total	Female	Total	Female
Total	85	170614	50033	120581	36708	9528	105960	32124	5731	1237	21064	6890	1151	254
Universities	40	151280	48274	103006	22560	8615	100913	31349	5603	1169	21064	6890	1140	251
Complexes	13	3305	1264	2131	1020	602	2295	628	80	34				
Colleges	4	2157	38	2119	127	3	2030	55						
Higher Ed. Schools	8	1308	247	1061	687	98	562	112	48	34			11	3
Technical Schools	20	12474	210	12264	12314	210	160	–						

Source: Research and Development Programs in the Five-Year Plan of the Islamic Republic of Iran, Ministry of Culture and Higher Education, 1988.

TABLE 4

NUMBER OF GRADUATES BY TYPE OF INSTITUTION,
LEVEL OF STUDY AND SEX, 1985–86

Type of Institution	No. of Institutions	Total			A.A.[1] Programmes		B.A. or B.S.[2] Programmes		M.A. or M.S.[3] Programmes		Professional[4] Doctorate		Ph.D[5] Programmes	
		Total	Female	Male	Total	Female	Total	Female	Total	Female	Total	Female	Total	Female
Total	85	28971	9356	19615	7586	2547	19346	6241	637	119	1221	416	181	33
Universities	40	26552	9136	17416	5835	2439	18689	6132	627	116	1221	416	180	33
Complexes	13	554	169	386	205	58	341	108	8	3				
Colleges	4	304	3	301	24	2	278	1	2					
Higher Ed. Schools	8	219	39	180	180	39	38	–						
Technical Schools	20	1342	9	1333	1342	9								

Source: Research and Development Programs in the Five-Year Plan of the Islamic Republic of Iran, Ministry of Culture and Higher Education, 1988

FIGURE 1
R&D EXPENDITURE AS A PERCENTAGE OF GNP

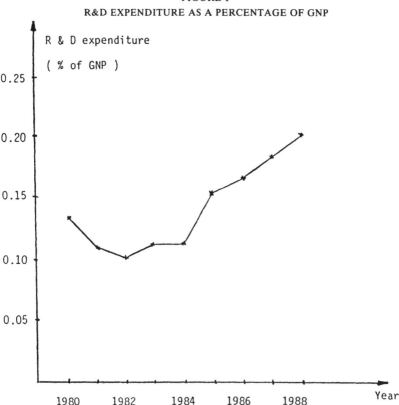

Source: Research and Development Programs in the Five-Year Plan of the Islamic Republic of Iran, Ministry of Culture and Higher Education, 1988.

It has been shown that R&D activities in terms of both the expenditure and number of researchers have been totally inadequate in Iran and steps have been proposed to increase these activities considerably during the five-year plan.

The government proposed a period of steady increase in the level of R&D expenditure as a percentage of GNP, so that at the end of the plan it reaches a level of one per cent. The Iranian parliament has initially reduced this to 0.6 per cent and intense debate is still going on to get the parliament's approval for the original target of one per cent. This being so, it is envisaged that the total number of full-time staff engaged in R&D would reach a level of 15,000 at the end of the five-year plan (a six-fold increase in five years).

TABLE 5

R&D EXPENDITURE AS A PERCENTAGE OF GNP (1985)

Country	R & D expenditure (% of GNP)
Portugal	0.42
Ireland	0.83
Italy	1.3
Norway	1.9
U.S.A	2.8
Japan	2.8
India	1.4
Brazil	0.94
Argentina	0.72
Iran	0.15

Source: *Research and Development Programs in the Five-Year Plan of the Islamic Republic of Iran*, Ministry of Culture and Higher Education, 1988.

TABLE 6

PER CENT DISTRIBUTION OF R&D EXPENDITURE IN IRAN (1979–88)

Type of Research	1979	1980	1981	1982	1983	1984	1985	1986	1987	1988
Industrial	62	64	65	66	74	71	70	70	76	73
Defence	-	-	2.1	4.0	2.6	2.0	2.2	2.7	3.1	2.7
Social	35	31	28	28	22	25	26	25	19	22
Others	0.6	0.2	4.9	2.0	1.4	2.0	1.8	2.3	1.9	2.3

Source: *Research and Development Programs in the Five-Year Plan of the Islamic Republic of Iran*, Ministry of Culture and Higher Education, 1988.

Details of the government proposals on the distribution of the R&D budget within the competing sectors are given in Table 7.

The general policy proposal regarding R&D activities in the five-year plan includes:

TABLE 7
DISTRIBUTION OF R&D EXPENDITURE (AS A PERCENTAGE OF GNP) DURING
THE FIVE-YEAR PLAN IN IRAN

Type of Research	1988	1989	1990	1991	1992	1993
University	0.026	0.048	0.070	0.092	0.114	0.136
Defence	0.006	0.042	0.078	0.113	0.144	0.185
Industrial	0.073	0.115	0.156	0.198	0.239	0.281
Agricultural	0.082	0.088	0.094	0.100	0.106	0.112
Social	0.023	0.038	0.052	0.067	0.081	0.096
Others	0	0.038	0.076	0.113	0.152	0.190
Total	0.210	0.369	0.526	0.683	0.841	1.000

Source: Research and Development Programs in the Five-Year Plan of the Islamic
Republic of Iran, Ministry of Culture and Higher Education, 1988.

(a) modifications in the current administrative procedures at the Higher
Council of Research, in order to simplify the process of starting new
research projects and for better co-ordination;
(b) setting up of various special committees at the Higher Council of
Research;
(c) establishment of research departments at various ministries with the
aim of facilitating the smooth flow of information to and from universities and other research institutions;
(d) consideration of the proposed research projects by the relevant special committees for approval.

Conclusions

The importance of the provision of manpower required for the reconstruction of the Iranian industry has been acknowledged by the authorities. Manpower requirement is looked upon as an important consideration
without which the successful completion of the five-year plan would be
hindered.

To ensure that a sufficiently educated and skilled work force exists to
facilitate reconstruction of the industrial sector of the economy, considerable efforts are being put by the authorities into education as a
whole. This is reflected by 7.5 per cent of GNP being spent on education
in Iran, a level higher than in Japan, Canada and most other developed
countries.

Activities in the research and development fields are much more limited and both research expenditure as a percentage of GNP and the number of researchers per million of population in Iran are significantly lower than most developing countries. To reverse this trend, drastic increases in the total expenditure on research and development have been proposed by the government and approved in principle by the Iranian parliament. It is envisaged that at the end of the present five-year plan, the number of full-time researchers per million of population will increase from the present level of 82 to about 300.

NOTE

The authors wish to thank the officials at the Ministry of Culture and Higher Education, Centre for Educational Planning and the Office of Vice-Chancellor for Academic Affairs, Ferdowsi University of Mashad for providing valuable information and statistics partly presented in this article.

REFERENCES

The President's speech at the third kharazmi ceremony for the prize winners in scientific research and innovations, Tehran, February 1990.
Research and Development Programs in the Five-Year Plan of the Islamic Republic of Iran, Office of Deputy Minister on Research and Development, Ministry of Culture and Higher Education, 1988.
Statistics of Higher Education in Iran, 1986–87, Ministry of Culture and Higher Education, Center for Educational Planning, Tehran, 1988.
UNESCO Statistical Digest 1987, UNESCO Publication.
World Bank Atlas 1987, World Bank Publication.
World Development Report, published for the world Bank by Oxford University Press, 1988.
World Military & Social Expenditures 1987–88, World Bank Publication.

EFFECTIVENESS OF
TECHNOLOGY TRANSFER

Technology Transfer from North to South

GEORGE McROBIE

The continued transfer of inappropriate technologies – large-scale and capital-intensive – from North to South can only exacerbate the South's problems. What is needed is a major expansion of the work of developing and helping to introduce appropriate technologies – small, simple, capital-saving and non-violent – of the kind described in this paper. The technologies of the North are also inherently unsustainable, and require drastic revision.

Now as never before, people in both the North and the South are becoming aware that the choice of technology is the most important decision that confronts national, regional and local governments, and others who determine what is produced, how it is produced, by whom and for whom.

For most of the past 30 years, indeed until quite recently, the prevailing notion of economic development was that cheap and abundant energy and rapid technological progress had opened up an era of limitless and potentially world-wide economic growth. Few questioned the survival value of an economic system based squarely on the heedless exploitation of people, the environment, and the world's stock of non-renewable resources, or doubted the wisdom of advocating its adoption by the then newly independent countries of the Third World.

Small is Beautiful

It was in relation to the needs and resources of the developing world that the deficiencies of rich-country technologies first became evident. The critical role of technology in economic development was first brought into

The author, Dr G. McRobie is the former head and the co-founder of the Intermediate Technology Development Group, Rugby, UK.

focus by E. F. Schumacher in the early 1960s. He argued that Third World countries were relying on rich-country technologies at their peril: that the large-scale, capital- and energy-intensive industries of the rich countries would do more to exacerbate than to solve the problems of the poor countries. Such technologies were singularly inappropriate because they:

- offer relatively few, very expensive workplaces, whereas the poor countries, with their masses of un- and under-employed, desperately need very large numbers of relatively inexpensive workplaces;
- are located chiefly in cities, which offer the mass markets scarce skills and infrastructure facilities not found in rural areas where the majority of the poor live;
- in many instances, compete out of existence traditional non-farm activities formerly carried on in rural areas;
- accelerate the migration of people from rural areas to metropolitan centres;
- make the developing country increasingly dependent upon rich countries for spare parts, skills and often markets;
- distort the cultures, as well as the economies of poor countries by concentrating economic activity in cities and social élites, breaking down rural structures (technology is not culturally neutral).

In 1965 a group of us helped Schumacher to start the Intermediate Technology Development Group in London. Our starting point was that mass unemployment and rural misery could be overcome only by creating new workplaces in the rural areas themselves; that these workplaces must be low-cost so that they can be created in very large numbers without calling for impossible levels of savings or imports; that produc-tion methods and associated services must be kept relatively simple, and that production should be largely from local materials for local use.

While it was (and still is) conventional in the field of development to put quite undue emphasis on GNP as a measure of success, the Group was informed by the knowledge that development basically means the development of people – their education, organisation and discipline, and their access to and control of the tools and equipment with which they can work themselves out of their poverty.

Our purpose was to demonstrate that technologies appropriate to the needs and resources of the rural poor could be developed and introduced, and then, by helping to create an international network of like-minded organisations, to change the whole emphasis of aid and development towards small-scale technology really capable of bringing industry into the rural areas.

Small is Possible

At first, and for several years, the Group did not get a very warm welcome either in rich or in poor countries. But then the conventional strategy of development, based on large-scale capital intensive industries, came to be increasingly challenged by development economists and planners. Many of the large industries proved to be very inefficient, kept going only by protection and subsidies. They did not generate the hoped-for surpluses and they did nothing to raise the living standards of the majority, the rural and urban poor.

By the mid-1970s the accumulating evidence of the failure of the large-scale industry strategy was accompanied by the dawning recognition that small-scale, localised industry and agriculture can reduce transport costs, decelerate city growth, produce goods and services very efficiently, and are the best way of distributing incomes. Then came the failure of the African agriculture; the vast and unrepayable Third World debt; and the relentless growth of unemployment in developing countries. These, largely man-made, disastrous developments, served to underscore the fact that encouraging the poor to behave as if they were already very rich only compounds their problems.

The direct transplanting of rich-country technologies into the South has already done much damage to the interests of the poor. Mushrooming cities (whose growth is closely associated with cheap oil) continue to grow apace. On UN projections, Mexico City, Sao Paolo, Calcutta, Cairo and Jakarta will all have more than 15 million inhabitants by the end of the century, and some of these will have between 20 and 30 millions. While in 1950 there were only six cities with more than five million, within the next decade there are likely to be no fewer than 60. The prospects in terms of energy and food supplies alone are daunting.

Employment prospects are no better. 'During the next ten years another 1.2 billion will enter working age . . . About a quarter of them will find some form of work in agriculture, industry or services. The remaining 900 million people will be unable to find a regular source of income . . .'.[1] This is surely a prescription for misery and violence on an unprecedented scale.

In terms of technology transfer, it is impossible to evade the conclusion that the attempt to transplant Northern technologies into the South has been in general a disastrous failure.

But it could be (and is) argued that the developing countries demand Northern lifestyles, and therefore Northern technologies and the kind of economic growth they bring. So is not the task that of adapting Northern technologies, so that they can be absorbed more gradually, equitably and efficiently? On this view, small-scale, low-cost appropriate technologies are not really an alternative but simply a stepping stone to the conventional technologies and lifestyles of the North.

The Economics of Impermanence

I am certainly not alone in believing this to be a dangerous misconception. In their predominant form, Northern technologies and their associated institutions are not sustainable. This is true both of industry and agriculture.

There is, first, the virtually total dependence of the North upon oil, which has decisively ceased to offer a low-cost, reliable or long-term energy supply; cheap oil has in fact proved to be an environmental disaster.

Second, conventional industrialisation is on a collision course with the environment. We now see the pollution of groundwater, air and food by industrial and agricultural chemicals; and the appalling prospect of living, if that is the right word, with what are to all intents and purposes permanently lethal radio-active wastes; the destruction of forests and the erosion of arable land; and the certainty of permanent climatic change arising from the greenhouse effect. Sooner rather than later we shall have to stop burning hydrocarbons in the atmosphere.

There are, above all, the human consequences of large-scale and highly centralised technology. The alienation and de-skilling of working people by mass production, the substitution of capital and energy for human skill – these have long been recognised but ignored in the interests of economic growth. But unemployment is now haunting both Europe and North America, and unemployment and alienation will continue so long as we treat labour merely as a cost. The growing centralisation of economic power also poses a threat to democratic processes.

Three Questions about Technological Development

We can no longer assume, in short, that conventional technology is appropriate for the purposes of environmental protection, the responsible stewardship of scarce resources, or the human need for useful and satisfying work. We must increasingly find ways of asking and demanding answers to these questions about technological development:

> What does it do to the resource base, renewable and non-renewable?
> What does it do to the environment?
> What are its social and political implications?

Effect on Resource Base

Modern large-scale industry has hardly proved to be a responsible steward of human or physical resources. Fortunately for us and for future generations, during the past ten years or so a growing number of organisations have started to address these questions. In Britain these include such excellent bodies as Friends of the Earth, the International Institute for Environment and Development, the Green Party, Greenpeace, the World-Wide Fund and the New Economics Foundation. Sustainable agriculture, unadulterated food and a healthy rural environment are the concern of bodies such as the Soil Association, the Henry Doubleday Research Association, and the London Food Commission. Largely thanks to such organisations, the actions required to create a more sustainable industry and society are now fairly obvious. These include:

- a determined programme of energy conservation; the elimination of nuclear power; the development of renewable energy sources;
- the progressive introduction of product standards and specifications leading to long-life products which can be readily repaired, renewed and recycled;
- a transport policy that rapidly diminishes the damage done by the internal combustion engine; and the promotion of public transport, especially rail;
- a rapid transition towards organic (non-chemical) agriculture;
- the location of economic activity, and the promotion of democratic forms of ownership and control, for example, by workers, communities.

The technologies that would emerge from such policies, and the values that inform them, would be very different from those which dominate the North – and which hold out no long-term future for anyone, anywhere.

Need for Ecological Approach to Agriculture

A particularly striking example of an unsustainable Northern technology is provided by agriculture. Conventional, chemical, agriculture causes damage to soil structure and increases soil erosion; damages the environment by polluting air, plants and water; adversely affects food quality and health; relies largely on fossil energy; and promotes ethnically unacceptable livestock systems.

Chemical farming is now rapidly spreading in the South. Although developing countries did not start using chemicals on any scale until around 1970, in the next decade their use of fertilisers grew twice as fast as in the Northern industrial countries. The FAO forecast future growth to the end of the century at about eight per cent per year. And by the mid-1980s the developing countries accounted for between 25 and 30 per cent of the world sales of pesticides.

It is tragic that, just as a growing body of opinion in the rich North is starting to demand that agriculture takes a different route towards sustainable, ecologically-based farming systems, and uncontaminated food and water, the poor countries of the world are increasingly being sold on the idea that chemical farming is the answer to their food problems. The prospect for poor countries was summed up by Dr Swaminathan, one of the world's acknowledged experts on food production, in an address to the Royal Society of Arts in November 1987. Having made several references to the need to preserve biological systems and the danger of going for short term gains at the expense of sustainability, he concluded: 'The global food scenario is one of hope on the production front. However, it is one of despair both in the field of equitable distribution and sustainability of the production pathways adopted . . .'.

The need for an ecological approach to agriculture in developing countries is spelt out by Nic Lampkin of the University of Aberystwyth, in a forthcoming book.[2] He refers to the 1989 conference of the International Federation of Organic Farming Movements (IFDAM), held in Burkino Faso, West Africa, which among its conclusions stated that ecological agriculture in developing countries was not an alternative, but a necessity imposed by local conditions; and that it is essential to develop national and international networks on agro-ecological systems of food production.

In short, chemical agriculture is a singularly inappropriate technology for developing countries. The appropriate technology is biological husbandry, an agricultural system based on ecological principles in which external inputs are minimised, in which chemicals such as fertilisers, pesticides and herbicides are not used, and which treat the soil as a living organism to be nurtured and kept healthy.

The chemicalisation of Third World agriculture can be stemmed and reversed only by identifying and making widely known, to all the organisations concerned, the principles and practice of ecological agriculture, including a body of case studies from different regions, and access to centres of research and training, and the relevant literature. At present there is no systematic, easily available body of knowledge of this kind.[3]

Identifying Needs: Collaboration Not Transfer

I have already indicated my conviction that appropriate technology (AT) offers the only feasible approach to development; and in what follows I try to highlight some of the salient featurs of AT viewed as an alternative strategy of development.

Because most poor people in the world make a living by working on small farms, in small family businesses or as artisans, technologies appropriate to their needs and resources will be small, relatively simple, inexpensive and (to be sustainable) non-violent towards people and the environment. But experience has shown that it is not by any means enough to produce and field-test such technologies. Devising or adapting the right hardware is part of a package which includes identifying the specific needs and resources of the community; developing a technology that can meet their needs – that raises their real income on a sustainable basis; and getting the technology widely introduced under operating conditions. Obviously, to be appropriate the technology should be capable of local operation and maintenance, and local or at least indigenous manufacture; it should be owned and operated by its users, and result in a significant increase in their net (real or money) income; it should utilise to the maximum extent local and renewable raw materials and energy; and it should lend itself to widespread reproduction using indigenous resources and through the medium of local markets.

An International Network

There are, today, at least 20 AT organisations around the world with a significant capacity for identifying needs and undertaking practical research and development work. In the industrialised countries there are the pioneer organisations, ITDG, VITA, Brace Research Institute, more recently joined by AT International, TOOL in Holland, GATE in Germany, GRET in France and IDRC in Canada. Among interna-tional organisations, UNICEF, the ILO, UNIFEM and IFAD are major supporters of AT work in developing countries.[4]

Anyone who doubts the vitality of the AT movement should reflect on the fact that 25 years ago the Intermediate Technology Group consisted of two of us in a small office in Covent Garden. Today the Group comprises more than 170 people, a highly professional staff of engineers and social scientists working on programmes in agriculture and fisheries, mineral industries and shelter, rural workshops and manufacturing, and renewable energy, in nine countries. The Group's American counterpart, AT International, specialises in rural small industry development; in the four years to 1988, some 1,200 small and micro enterprises were set up with ATI's help in 22 countries, the cost per workplace averaging US$1500.

Within the developing countries there are now hundreds of AT organisations, ranging from technical R&D groups at one end of the spectrum to information-networking teams at the other.

Choosing the Right Technology

Thus today, over a wide range of human activities, especially those related to basic human needs, technology choices are becoming available. Small-scale, low-cost technologies now exist in agricultural equipment and food processing, water supply, building materials, textiles, small-scale manufacturing, energy, transport. What is now beyond question is that technology choices can be created for all practical purposes across the board. When high quality engineers turn their minds to devising small-scale, capital and energy-saving technologies, they can produce some remarkable results.[5]

More R&D Needed

Yet many gaps in knowledge remain, and the funding of R&D in appropriate technology is derisory in relation to the need. For the developing countries the creation of hundreds of millions of new workplaces in agriculture and industry during the next couple of decades is now an overwhelming necessity. It should be one of the most urgent tasks on the agenda of aid-giving governments and international agencies to ensure that at the very least the technologies enabling a basic needs strategy to be implemented should be readily available to the governments and people of developing countries. This calls for a major expansion of R&D and field testing, and publication of data on, appropriate technologies. The impending energy crisis in many poor countries alone requires much additional work on these lines on renewable energy. And every developing country should possess an indigenous AT organisation capable of advising government of the technology choices available and of the economic and social implications of different choices, advising on the implementation of AT programmes, and undertaking R&D and technology adaptation to local conditions.

Conditions of Success

As yet, relatively few ATs have spread widely, although many have achieved local success. Appropriate Technology International has recently compiled the first of a series of detailed case studies entitled 'High-Impact Appropriate Technologies',[6] which include these examples:

Mark II handpumps in India: The Mark II deepwell handpump (18 to

50m.) is now the basis of widespread community water supply in India. More than 600,000 are now installed, serving 120–150 million villagers. Some 38 firms employing 8,500 people are now engaged in making these pumps at the rate of about 156,000 a year. Some 50,000 people are employed in well drilling or maintenance of existing installations. At village level there is a pump caretaker equipped with a set of tools; at sub-district level, there are roving mechanics; at district level, mobile maintenance teams. The per capita cost to the users (covering the well, pump and maintenance costs) is less than 1 dollar a year. During the past five years, an estimated 15,000–20,000 pumps have been exported.

Implementing organisations are UNICEF, Government of India, State Governments, villages and communities, private and public pump manufacturers.

Oral rehydration therapy: This is a simple, inexpensive and effective way of treating diarrhoea in young children. Over the past 15 years ORT has spread to virtually every developing country, and aboout 100 million packets of ORT sales are produced and distributed annually. There is a growing number of local production units in developing countries, and a home preparation of a basic unit is now envisaged.

Implementing agencies are UNICEF, Red Cross, WHO, Government and non-Government health services, communities and families.

Water pumping windmills in Argentina: About 60,000 water-pumping windmills are currently in use in Argentina and annual production is 1,500 to 2,000. About 300,000 people benefit by getting water for their livestock and own use.

Implementing agencies are private manufacturing companies.

Bamboo-reinforced concrete water tanks, Thailand: More than 24,000 tanks have been installed since 1979. Villagers contribute their labour, pay materials costs and a small surcharge into a revolving loan fund. Up to 10,000 more tanks are likely to be built in the next three years.

Implementing agencies are Community Based Appropriate Technology and Development Services; some funding from ATI, Ford Foundation, IDRC and CUSO.

Bamboo tubewells, India: Work on bamboo tubewells started in 1967. By 1980 there were some 100,000 wells in Bihar and Uttar Pradesh. A mobile pump services five or six tubewells, which operate at depths of 30/36 metres; work is now in hand on a bullock powered pump. The technology was developed by farmers and small contractors. Its cost is one-third to one-half of that of steel tubewells.

Implementing agencies are Deen Dayal Research Institute, Governments of Bihar and UP, local entrepreneurs.

Rural access roads programme, Kenya: An example of the AT approach applied to civil engineering. Competitive with capital-intensive methods, the programme established more than 40 field units and completed some 7,000 kilometres of rural roads. It incorporates a technical service unit and a training programme, and employs about 8,000 labourers.

Implementing agencies are Ministry of Transport, Kenya, IBRD, ILO and UK and other bilateral donors.

Women's pappad processing co-operative enterprises, India: Started 25 years ago when seven women invested 80 rupees in a low-cost nutritional snack food venture. Today the business generates sales of 30 million rupees and provides income for more than 6,000 co-op members. Raw materials are bought in bulk and distributed to the co-operative's 21 centres, which operate with a good deal of autonomy, and are run entirely by women.

Implementing agency is Lijjat Pappad Women's Co-operative.

Rural small farm implements components manufacture, Tanzania: A decentralised approach to the production of ox-drawn implements (ox-carts and toolbars) for small farmers. Two small factories are in production and two others are planned. In two years more than 2,000 ox-carts and 2,500 toolbars have been sold. The equipment pays for itself in a matter of months rather than years, as it enables more land to be cultivated. Surveys indicate that only a small fraction of the demand for this equipment has yet been met.

Implementing agencies are Tanzanian public and private implement manufacturers and USAID.

Other projects with similar potential are not hard to find. Some of the more striking examples that I have come across recently include the local manufacture of *fibre-reinforced roofing tiles*, an ITDG project in Kenya. From a pilot project of ten production units this is envisaged as 50 production units after four years. By then, local income generated from the project would amount to two million Kenyan sh. annually. Low-cost, local manufacture of FCR roofing tiles has potential in practically every developing country. Another example is the small-scale, locally-made *sorghum and millet* dehuller, developed and funded by the International Development Research Centre in Canada in collaboration with several African Countries. Some 35 dehullers are in commercial operation in Botswana, 40 are planned in Zimbabwe, 10 are working in the Dominican Republic, and pilot schemes are starting in India and Gambia.

A programme which is well past the pilot stage is ITDG's introduction of *plywood fishing boats* in South India. At least 2,000 fishermen are earning a good living from 400 plywood boats operating along the south-west coast of India. There is now a firm base for a decentralised, capital-saving, skill and labour-using industry there, for boat-building and repairs. (It should be added here that it is far better to 'unroll' a huge mango tree into plywood which can make many boats, than to use the same tree to make one dugout canoe and a heap of wood chips!). To cite one of the projects of AT International, the introduction of a *small oil press and associated equipment* to small farmers in Cameroon is likely to raise their incomes by at least 50 per cent. This project demonstrates the 'package' character of successful rural projects. It includes a locally-controlled revolving fund which both finances the production of the equipments and enables groups of local farmers to buy it. The farmers first lease the equipment, then purchase it out of their increased incomes.

Reasons for Success

Why did the particular technologies outlined above succeed?

First, the technologies themselves have been thoroughly field-tested and refined before going into production – the history of handpumps failures is impressive, for example, and tens of thousands of man-hours have been spent on getting animal-drawn equipment right. The technologies lend themselves to local manufacture wholly or in part, and to local maintenance; and they are low cost enough to be afforded by individuals or working groups of the 'target' population.

Second, the users or beneficiaries of the tehnologies are closely associated with processes of selection, introduction and use of the technology or product; and from the standpoint of the users, the advantages (in the form of cash or higher real income or life enhancement) significantly outweigh the costs incurred by them.

Third, the technologies are disseminated through the mechanism of the market; but in practically every case the market alone would be inadequate – it has to be supplemented in several ways. Thus, R&D and testing are preconditions of a product making its appearance in the market. But neither its appearance, nor detailed information about it, puts it in the hands of the rural poor. The poor have no money, or none to spare; their needs must be translated into effective demand – that is, demand backed by purchasing power. This requires that the poor have access to credit enabling them to buy the new equipment. They can then, as in the case of the oil-press in Cameroon, repay the loan out of the increased incomes they get by using the new technology. Credit that enables the poor to become more productive lies at the very heart of the process of rural development, of capital accumulation and income generation in rural areas.

The Power of the Package

Careful attention must also be given to such matters as quality control, training, extension and demonstration, and the creation of new local institutions to ensure continuity of user control and benefit. In all instances, that is, the hardware part of the technology is part of a package which empowers local people to choose what suits them best; gives them access to a low-cost, good quality product over which they have a good measure of control; and which enables them, by using it, to raise their standard of living. Essentially this is investing in people by making them more productive.

One important reason for the widespread adoption of these technologies is that they have either had support at government level, or that they have in some way overcome or by-passed the formidable obstacles that inhibit ATs in most developing countries. These are familiar enough – development strategies based on top-down, large-scale programmes and projects, and financial policies, administrative procedures and rules that favour the big over the small, the urban over the rural, the rich over the poor. If such obstacles could be lowered or removed, the dista-strous consequences of the conventional large-scale technologies would be clearly revealed, and cost-effective ATs would be widely adopted through the market.

The importance of government policies in fostering or inhibiting rural development and small enterprise has been thoroughly explored by Frances Stewart and others[7] and recently elaborated with detailed case studies by Ton de Wilde (until recently head of AT International) and Stijn Schreurs.[8]

Conclusions

I conclude with these observations:

The transfer of technology, in the sense of transplanting the technologies of the rich countries mostly into the cities of the South, does little or nothing for the poor majority; and many of the problems of the South can be attributed to this kind of top-down process – which still continues through conventional programmes of aid and development.

Experience has shown that appropriate technologies, which enable the poor to work themselves out of their poverty, can be arrived at only by collaborating with local people in identifying their needs and resources; and by embodying the technology in a package which includes credit, materials supply, marketing, and technical training and back-up.

There is now a substantial and growing body of case studies and field experience, convincing evidence that AT is an efficient, cost-effective way of meeting a wide range of human needs. That relatively few ATs have spread widely has led to a recognition that national policies – fiscal, financial, administrative – very often work against the spread of appropriate technologies through the market, and that such policies should, and can be changed.

Both the development of ATs and their widespread use is likely to accelerate, with the growth in the number and expertise of voluntary organisations in the South. The growth and influence of voluntary organisations at local and national level in the South, representing people as producers, is one of the guarantees that appropriate technologies are developed, and that they remain under the control of the poor. The role of the voluntary organisation becomes more important, the more that economic and political power become centralised. As redistributors of economic power their role is of crucial importance, alike in countries of the South, and of the North.

NOTES

1. Ton de Wilde and S. Scheurs, *Opening the Market Place*(London: IT Publications, 1991).
2. *Organic Farming* (Ipswich: Farming Press, 1990).
3. The New Economics Foundation and the Soil Association propose to set up an information and documentation service on organic husbandry in developing countries.
4. A. Sinclair, 'A Guide to AT Institutions' (London: IT Publications, 1984).
5. M. Carr, 'The AT Reader' (London: IT Publications, 1985).
6. T. Fricke, 'High Impact Appropriate Technology Case Studies' (Washington, DC: AT International).
7. *Macro-Policies* (first published by Westview Press, Colorado in 1987) (London: IT Publications).
8. Ton de Wilde and S. Scheurs, op. cit.

Transfer of Technology to Less Developed Countries: A Case Study of the Fertiliser Industry in Bangladesh

M. M. HUQ AND K. M. N. ISLAM

The paper examines the growth of the urea fertiliser industry in Bangladesh in the context of the transfer of technology that has taken place. It appears that the technology transfer process has remained more or less confined to the acquisition of production capability; that project preparation and appraisal were generally inadequate, and that local participation and implementation in project design was nil or negligible. An important point emerging from the transfer process is the strong link between the source of machinery and equipment and the source of foreign aid, and the even stronger connection between the source of foreign aid and that of the general contractor.

Introduction

The main objective of the paper is to throw some light on the nature of technology transfer which has taken place in one high technology industry, the manufacture of fertiliser, in a Third World country, Bangladesh. In Bangladesh, investment in fertiliser has been very high – around three-quarters of the total public sector expenditure in manufacturing during the latter half of the 1980s; also the net value of the fixed asset gives the industry the top rank in the manufacturing sector in 1987–88, the last year for which such data are available. Technology for the industry has invariably been imported, thus providing a good case for studying the nature and extent of the technology transfer.

In order to enable us to view the transfer of technology in some detail we have concentrated our discussion on the manufacture of urea which forms the main output (85 per cent) of the Bangladesh fertiliser industry.

The paper has been arranged as follows: section I gives a brief introduction to the growth of the industry in Bangladesh; section II is in two

M. M. Huq and K. M. N. Islam are respectively Senior Lecturer in Economics, University of Strathclyde, Glasgow, and Research Associate, Industry Division, Bangladesh Institute of Development Studies, Dhaka.

sub-sections – the first throws some light on the development of fertiliser technology in the world, the second on the technology in use in Bangladesh. Section III analyses the transfer process from the recipient's viewpoint. Finally, in section IV, we draw some conclusions.

I. Growth of the Fertiliser Industry in Bangladesh

The first fertiliser plant in Bangladesh started commercial production in 1962. Since then, four other urea plants and one phosphorus plant have been built, and another urea plant is now under construction. The present total installed capacity of the operational plants is 1.7 million tonnes of urea (2.32 million tonnes when the plant under construction becomes operational).

Average annual growth of urea production is found to have been 11.1 per cent since the early 1980s. Urea remains the dominant fertiliser in the country. Available data on the use of fertilisers per unit of cropped land shows a significant increase since the mid-1970s, from 11 kg nutrients per hectare of land in 1969–70 to 40kg in 1985–86. However, the increase of fertiliser use is significantly associated with the increase in irrigated area. The proportion of land irrigated by mechanical methods to total cultivated land, which was less than 4 per cent in 1969–70, increased to about 20 per cent in 1985–86.[1]

At present, given the level of demand there is already excess installed capacity in the fertiliser sector in Bangladesh. The situation is particularly disturbing as, following a large drop in the export price of fertiliser, export is found to be a losing concern. Once the Jamuna Fertiliser Factory, which is being constructed, becomes operational in two years' time the situation will be worse if the export price does not improve and if there is no significant increase in domestic demand through expansion in irrigation and necessary land reforms.

Donors have been found particularly keen to lend money for the establishment of fertiliser plants and, in an aid-dependent development situation as in Bangladesh, the recipient is found to be very enthusiastic about offers of aid. It appears that, in the process, the likelihood of installed capacity remaining unutilised has not been very carefully examined.

II. The Use of Fertiliser Technology

Fertiliser Technology in the World

In the manufacture of urea, the initial stage is to produce ammonia by using natural gas. The Haber-Bosch process, invented in 1903, showed how to produce ammonia. The original Haber-Bosch process has witnessed significant improvements through minimising the use of feedstocks (for example, natural gas and various catalysts), thanks to research and development mainly carried out by some large companies. As a result, a number of companies now control economical process design, as may be seen from Table 1. These companies usually give licences to contractors wishing to establish ammonia plants. Similarly, there are also companies which have economical process designs for the manufacture of urea. At times, some of the companies owning such process designs – either in ammonia or in urea – take direct orders for the establishment of fertiliser plants. But it is also possible for others to buy licences from the companies which control the process designs.

Thus, we find that a few established companies have acquired the knowledge of the commercial manufacture of ammonia and urea. These companies control this knowledge in an oligopoly form, as characterised by Ghatak and Disney.[2]

As in the case of the supply of processes, there have also emerged some specialised companies which take general contracts for the establishment of fertiliser plants. These general contractors become responsible for the successful establishment of fertiliser plants. Table 1 also gives an idea of some general contractors and their country of origin, and also whether they possess any licence for ammonia urea. As may be seen from the table, these companies are mainly from the developed countries, Japan emerging as a very important source. From the Third World, only in India and China do we find companies working as general contractors.[3] As will be shown later, there is a strong connection between the source of machinery and equipment and the source of foreign aid. The connection between the source of aid and that of general contractors is even stronger.

Technology in Use in Bangladesh and Its Source

Table 2 has been prepared to give an idea of the sources of processes and of machinery and equipment obtained for the various nitrogenous fertiliser plants in Bangladesh. As may be seen, Japan has played a very important part in the development of the urea manufacture in Bangladesh, but there are also other sources including Western Europe, North America and China.

Multiplicity of sources both for licences for process design and the supply of machinery and equipment has remained an important feature in the production of ammonia and urea in Bangladesh. Information collected, however, suggests that attempts at cost minimisation were not

TABLE 1

SOURCES OF TECHNOLOGY: PROCESS DESIGN & GENERAL CONTRACTORS

a) Ammonia and Urea Processes: Main Licensers

	Country of origin of the licencer	Licences given to plants operating in Bangladesh
Ammonia Process		
ICI	UK	–
Kellogg	USA	Chittagong Urea Fertilisers Ltd.
Halder Topse	Denmark	Jamuna Fertiliser Co. Ltd.
Uhde	W. Germany	Zia Fertiliser Co. Ltd.
Urea Process		
Stamicarbon	Netherlands	Zia Fertiliser Co. Ltd.
TEC-MTC	Japan	Chittagong Urea Fertliser Ltd.
SNAM-progetti	Italy	Jamuna Fertliser Co. Ltd.

b) General Contractors

General contractors	Country of origin	Plants supplied to Bangladesh
Mitsubishi Heavy Industries	Japan	Jamuna Fertliser Co. Ltd.
Kobe Steel Ltd.	Japan	Natural Gas Fertiliser Co.
Toyo-Engineering Corporation (TEC)	Japan (Urea)	Urea Fertiliser Factory Ltd Chittagong Urea Fertiliser Ltd.
Friedrich Uhde Gub.	West Germany (Ammonia)	Zia Fertiliser Co. Ltd.
M.W. Kellogg Company	USA (Ammonia)	–
Chemical Construction Corp.	USA (Ammonia)	–
Kellogg International Corp.	UK	–
Montecatini	Italy	–
Foster Wheeler Ltd.	UK, USA	Zia Fertiliser Co. Ltd.
India Engineering Ltd.	India	–
Power & Dev. India Ltd.	India	–
CNCCC	China	Palash Urea Fertiliser Factory

Note: Production process, if any held by individual firms, is shown in brackets.

Source: Information obtained from fertiliser experts.

the sole factor which dictated the procurement of production processes or of the required machinery and equipment. In fact, the development of the industry has been largely dictated by the availability of foreign aid, and the restrictions which accompanied such aid often dictated the sources of procurement. At times even the appointment of the general contractor was in some way dictated by the aid giver. One therefore does not expect any serious search to achieve the procurement of least-cost production processes and machinery and equipment. The procurement process is further restricted by the small number of established licensers of processes and general contractors. It is true that in some cases, for example for the Jamuna Fertiliser Company Ltd (JFCL) the price was reported to be competitive,[4] but this cannot be said for some other plants. In fact, it is strongly believed by many that the Zia Fertiliser Company Ltd (ZFCL) and also the Chittagong Urea Ltd (CUFC) were not obtained at a competitive price; in other words, it was possible to have obtained these plants at a lower capital investment. It is true that cost comparisons of plants in different locations can be difficult because of the differences in circumstance and infrastructural facilities that exist according to locations. But these differences cannot fully explain the huge difference in capital expenditure. For example, in the case of the Chittagong plant, the total capital investment was about US$500 million while in the case of the Jamuna plant the total investment is less than US$350 million (fixed lump-sum payment); in both plants the scale of production is more or less the same. (At constant prices, the difference would be even larger considering that the Chittagong plant was built earlier.)

So far as machinery and equipment are concerned, these are usually bought by shopping around the world, although the degree of competition is likely to be limited because of the specialised nature of some of the items required. Moreover, only a handful of companies are pre-qualified to supply the required items (a normal practice in Bangladesh) and there is the possibility of some collusion between these suppliers.[5]

III. Technology Selection and Transfer

From the information we have been able to collect, it appears that the technology supply has been largely dictated by foreign suppliers, and this is particularly so for the earlier plants. For example, the decision for the establishment of the Natural Gas Fertiliser Factory (NGFF) at Fenchuganj was an outcome of interest shown by Kobe Steel of Japan. Kobe was at that time (1960) building a fertiliser plant for Japan and it proposed to the then Pakistan Industrial Development Corporation (PIDC) to establish a similar plant in Sylhet so as to utilise the natural gas

TABLE 2

SOURCES OF FOREIGN CURRENCY AND TECHNONOGY IN THE VARIOUS
NITROGENOUS FERTILISER PLANTS IN BANGLADESH

Plant Name	Source of foreign currency	Technology supply		Major sources of machinery & equipment	General contractor	
		Name of process	Country of origin		Name	Country of origin
NCFF	Japanese loan	*Ammonia*: Chemico *Urea*: Chemico Total recycle	USA USA	Japan	Kobe	Japan
UFFL	Yen credit	*Ammonia*: TEC *Urea*: Toyo Total Recycle	Japan Japan	Japan	Toyo Engg. Corp. (TEC)	Japan
ZFCL	IAC	*Ammonia*: UHDE, *Urea*: Stamicarbon	W. Germany Netherlands	W. Germany Italy, India, UK	Foster Wheeler Ltd.	UK
PUFF	China	*Ammonia*: Chinese process *Urea*: Chinese process	China China	China	CWCCC	China
CUFL	OECF, IDA CIDA, IsDB	*Ammonia*: Kellogg *Urea*: TEC-MTC	USA Japan	Japan, Canada, Germany	Toyo Engg. Corp. (TEC)	Japan
*JFCL	OECP Japan	*Ammonia*: Halder Topse *Urea*: SNAM-Progetti (Urea granulation: NSM)	Denmark Italy	N.A.	Mitsubishi	Japan

IAC = International aid consortium
NGFF = Natural Gas Fertiliser Factory.
UFFL = Urea Fertiliser Factory Ltd.
ZFCL = Zia Fertilier Company Ltd.
PUFF = Polash Urea Fertiliser Factory.
CUFL = Chittagong Urea Fertiliser Ltd.
JFCL = Jamuna Fertiliser Co. Ltd.
* The plant is under construction.

Source: Feasibility Studies of the various plants; *Project Completion Reports* and discussions with relevant persons.

which was discovered in that area. It is true that the technology for the manufacture of urea was then at an early stage of development and, one can say, there was no serious scope for evaluating various technologies at that time. Moreover, Pakistan in the late 1950s had not developed the expertise for evaluating alternative fertiliser technologies.[6]

The Urea Fertiliser Factory Ltd (UFFL) was built in 1972 at Ghorashal, which is a large-scale plant, over three times the size of NGFF. It was built by Toyo Engineering Corporation with Yen credit, provided by the Japan government. Production processes used for ammonia and urea were also obtained from Japan. Like NGFF, the Polash Urea Fertiliser Factory (PUFF) is of smaller scale and type and the investment decision for the project was made during a visit to China by President Ziaur Rahman more on political grounds than on any systematic economic evaluation. So one can say that no serious appraisal was made for selecting technology in this plant, which is termed a 'friendship project' between China and Bangladesh. In the case of the other three plants which are of the large-scale type, international tenders were invited and on the basis of the 'tender bids' received general contractors were appointed, though for the plant at Ashuganj (ZFCL) the appointment of the general contractor was reported to be largely pushed by the aid givers through the World Bank.[7]

Table 3 gives an idea of the nature of the selection process in the various fertiliser plants, producing urea in Bangladesh. Failure on the part of the recipients to have the production processes and the required machinery and equipment properly evaluated can obviously lead to bad investment decisions. Considering that the evaluation procedure in this respect for investment in the fertiliser sector of Bangladesh has not been as one would have desired, especially given the large capital investment involved, there is obviously room for criticism.

It appears that investment decisions were often taken by non-technical personnel, the participation of technocrats remaining peripheral or very limited. The need for taking decisions on the basis of advice from knowledgeable experts can hardly be over-emphasised. The technocrats are obviously well-placed to make informed decisions based on their knowledge of technological developments taking place in the industry. The Bangladesh Chemical Industries Corporation (BCIC) is the controlling body of the various fertiliser plants currently operating in Bangladesh and this Corporation has a Planning and Implementation cell which is responsible for taking decisions regarding the expansion and renovation/rehabilitation of fertiliser plants in the country. This cell which is manned by experienced fertiliser experts does, however, lack the required R & D back-up facility. A comparison can be made with the Indian process of decision-making, where the technocrats are generally asked to take the

TABLE 3

TENDERING, PROCESS EVALUATION AND PURCHASE OF MACHINERY AND
EQUIPMENT IN THE VARIOUS NITROGENOUS FERTILISER PLANTS

Name of the plant	Year of Initiation Project	International tenders called for the project	Processes separately evaluated by the recipient	Selection of M/C and equipments			Any agreement to procure local M/C & equipment	
				Tendering	Evaluation by source of procurement	Evaluation by engineering type	Yes/No	% of Local Purchase
NGFF	1960	Yes	No	Yes	No	No	No	Nil
UFFL	1969	Yes	No	Yes	No	No	No	Neg.
ZFCL	1975	Yes	Neg.	Yes	No	Partly	No	Neg.
PUFF	1982	No	No	No	No	No	No	10
CUFL	1986	Yes	Yes	Yes	Implicit	Partly	No	Neg.
JFCL	1988	Yes	Yes	Yes	No	N.A.	No	N.A.

Neg. = Negligible
Source: Field Survey

final decision and they are often helped by specialised research centres of high calibre.[8]

As Table 3 shows, the technology selection-mechanism in the case of the fertiliser plants in Bangladesh can be considered far from perfect. Information available for the relatively new plants shows that the selection procedure is often confined to the appointment of general contractors for the establishment of individual plants. The production processes, specified by the general contract in the (tender) bid, are automatically selected.

As the acceptance of bids is mainly on the basic price quoted it is assumed that the bidder (who is one of those already pre-qualified) asking the lowest price is offering the best technology. In the case of 'lump-sum' contracts, the selection of machinery and equipment is the responsibility of the general contractor; but in the case of 'cost-plus-fees' contracts, the recipient decides, in consultation with the general contractor, the machinery and equipment. However, in the case of the Ashuganj Plant, the general contractor, Foster Wheeler (UK) Ltd., initially was very powerful and the recipient's views were either not sought or given little importance.[9] At times, the recipient failed to give comments within the specified time, thus providing the general contractor with an excuse to go ahead with the orders of its liking.

In the case of the CUFL plant, which was also obtained on a 'cost-plus-fees' basis, the General Contractor – Toyo Engineering Corporation (TEC) – was, however, able to establish a good working relationship with the recipient and the regular consultation, as required, continued without

serious difficulties. The TEC was also found to be very receptive to the
well-reasoned arguments put forward by the Bangladeshi experts.

Table 4 gives an idea of the participation of Bangladeshis in plant
design, erection and supervision. As may be seen, the achievement on all
the counts is very poor. The only exception is the PUFF plant, supplied by
the Chinese (in the form of a friendship project), for which both the
supplier and the recipient government wanted active local participation
in, at least, supervisory works.

As regards plant design, absolutely no transfer of technology has taken
place in any of the plants, and almost the same thing applies concerning
plant erection. The low or negligible involvement of the local engineering
sector in all the plants can also be severely criticised.

For the sake of argument one can advance two main reasons for the
non-use of local machinery and equipment in industrial plants established
in a country like Bangladesh. First, the technology is of such a composite
type that the entire machinery and equipment is fabricated/produced in a
high-level engineering plant in advanced countries. Second, the type
of materials required is not available in a developing country such as
Bangladesh. The above arguments, however, do not hold as far as the
fertiliser plants are concerned. Indeed, according to a recent study by
the Bangladesh Steel and Engineering Corporation (BSEC) the local
involvement in terms of structural steel works, storage tanks and pressure
vessels, including piping, could be very high.[10] For example, manufacturing of atmospheric vessels and low pressure vessels of carbon steel
(requiring no stress annealing) is possible locally using imported raw
materials. This represents about one-tenth of the total cost of a typical
large-scale fertiliser plant. According to the BSEC study mentioned
above, as much as 37 per cent of the total project work of a large-scale
fertiliser plant can be completed locally.[11]

Concluding Remarks

With the rapid expansion of the fertiliser industry in Bangladesh over the
last three decades the country has acquired production capability in the
industry. However, for a genuine transfer of technology a country should
also acquire investment capability in the form of plant design, machinery
manufacture, and even some innovation and invention – all these to
follow from the initial achievement of production capability.[12] In the case
of the Bangladesh fertiliser industry, the country has remained totally
dependent on the entire import of the production process, plant design
and machinery and equipment. Even the supervision of plant construc-

TABLE 4

LOCAL PARTICIPATION IN DESIGN AND SUPERVISION OF PROCESSES AND
PLANTS

	Plant Design	Plant Erection	Local Fabrication	Supervision of Plant Construction
NGFF	Nil	Nil	Neg	Nil
UFFL	Nil	Nil	Neg	Nil
ZFCL	Nil	Nil	Nil	Nil
PUFF	Nil	Neg	10%	30%
CUFL	Nil	Neg	Neg	Nil
JFCL	Nil	-	-	Yet to be constructed

Neg = Negligible
Source: Factory visits and interviews with people who originally participated in
the establishment of the plants.

tion and also the installation of various machinery and equipment have
been almost entirely carried out by the foreign contractors.

A number of factors can be identified for the poor state of technology
transfer in the fertiliser industry of Bangladesh:

1. The aid-dependent nature of Bangladesh's economic development
   has implied a total dependence on foreign sources for the wholesale
   import of technologies. The conditionalities which go with foreign aid
   have obviously not helped in involving the Bangladeshi experts at the
   various stages of technology transfer.
2. As there has not been any insistence on the part of the recipient
   for using local machinery and equipment and for local fabrication,
   the general contractors found it easier not to make use of the local
   resources. Indeed, there has not been any conscious and systematic
   government policy to this effect.
3. For the local management staff and the engineering personnel to
   acquire relevant knowledge in the field it is necessary that they are
   involved in the setting up of the industrial plants. It will be particularly
   necessary to support them when they fail and the process of learning
   by 'trial and error' should be accepted as national policy. But, such a
   policy is yet to emerge in Bangladesh.
4. R&D is an essential part of the process of acquiring indigenous technology capability and, in industries such as fertiliser production where
   basic research is not normally carried out in Bangladeshi academic
   institutions, it is essential that the fertiliser sector initiates such R&D
   activities effectively. It would be wrong to expect an immediate return

from the R&D expenditure. A country such as Bangladesh has much to learn, especially from the Indian and Chinese attempts in this regard.[13]

In brief, genuine technology transfer will not easily take place in an industry such as fertiliser manufacturing, unless there is a systematic and conscious national policy, especially when the country is heavily dependent on foreign aid for its economic development.

NOTES

The paper is based on the findings of the research conducted in July–September 1989 at the Bangladesh Institute of Development Studies, Dhaka with research support from the UNDP and the Ford Foundation. See M.M. Huq and K.M.N. Islam, *Production of Fertilizers in Bangladesh: An Industry Study with Particular Reference to Choice of Technology*, BIDS, Dhaka, 1990. The authors are grateful to an anonymous referee and also to Drs P. Bhatt and A. Shibli for helpful comments.

1. M. Hossain, 'Fertilizer Consumption, Pricing and Foodgrain Production in Bangladesh' in Bruce Stone (ed.), *Fertilizer Pricing Policy in Bangladesh* (1987).
2. S. Ghatak, *Technology Transfer to Developing Countries: The Case of Fertilizer Industry* (1981) and R. Disney 'Scale and Efficiency in the Chinese Nitrogenous Fertilizer Industry', *Indian Economic Review* (1979).
3. The plant design capability of companies from India and China is, however, found to be limited and they usually seek help from outside.
4. According to a representative of Mitsubishi Industries (the supplier), they could offer low prices because they themselves are the manufacturer of machinery and equipment. However, according to another source, the heavy machinery industry sector was in a depression when the tender was invited and Mitsubishi was keen to get the order.
5. Only those companies which according to the recipient are competent to tender will be entitled to do so. In other words, according to the terminology used only the pre-qualified companies can submit tenders.
6. At that time, another fertiliser plant was set up at Mardan, West Pakistan, with technology imported from France and the decision to establish it was reported to have been made by the then Chairman of PIDC, Mr Golam Farooq, who was known to have taken the decision without even doing the basic paper work. The plant turned out to be a very bad one, and ultimately had to be scrapped.
7. Iqbal Mahmud, 'Technical Management of Aided Project: Study of a Large Scale Process Industry in Bangladesh' (1984).
8. See Ch. 2 in Asian and Pacific Centre for Transfer of Technology (APCTT), *Technology Policies and Planning: India* (Bangalore, 1986) for some useful information on India's recognition and commitment to technology for development.
9. Iqbal Mahmud, op cit.
10. Bangladesh Steel and Engineering Corporation (BSEC), *Proposal for Local Participation in the Construction of Jamuna Fertilizer Projects and Karnafully Fertiliser Project*, Dhaka 1989 (mimeo).
11. Ibid., p. 4-1. The point refers to Jamuna and Karnafully fertiliser projects.
12. See L.E. Westphal, L.-S. Kim and C.J. Dehlman, 'Reflections on Korea's Acquisition of Technological Capability', World Bank, Economics and Staff Research Paper, April 1984. See also J.L. Enos and W.-H. Park, *The Adoption and Diffusion of Imported Technology* (London: Croom Helm, 1988).
13. See Asian and Pacific Centre for Transfer of Technology, *Technology and Policies Planning Study: India* (1986) and *Technology and Policies Planning Study: People's Republic of China* (1986).

Appraising Small-Scale Modern Dairy Development in Kenya

MICHAEL TRIBE

The paper considers technology and scale choice in milk process-
ing in Kenya – specifically in an area remote from large-scale
processing facilities. Raw material deterioration is of paramount
concern, together with insertion of modern small-scale technology
into the established market structure. UHT (ultra high tempera-
ture) treatment significantly increases the shelf life of processed
milk over raw milk, but 'diseconomies of small-scale' associated
with miniaturising the modern technology endangers economic
viability by comparison with larger scale plants using the same
technology type. The small-scale modern technology is therefore
in danger of 'falling between two stools'.

Introduction

For some years the Government of Kenya (GoK) has encouraged the development of small-scale dairy processing plants in rural areas which are too remote to be able to send raw milk on a regular basis to the large-scale dairies of the Kenya Cooperative Creameries (KCC).[1] The pilot projects commissioned in 1985/86 in Meru and Bungoma Districts are of interest as examples of the application of modern technology to rural development:

(1) The encouragement of small dairy plants is consistent with the policy of 'District Focus Planning' which has been given so much emphasis in GoK statements in recent years, with 'rural industrialisation' and with off-farm employment creation.[2]
(2) Dairy development is relevant to better nutritional standards in rural areas through enlargement and improvement of dairy stock, and improved access to dairy products for the population with better distribution systems and changed product and packing characteristics: an aspect of the 'Basic Needs' approach to economic development.[3]

Michael Tribe is a Lecturer in the Development and Project Planning Centre, University of Bradford, England.

187

(3) Dairy development can increase farm incomes, particularly in areas of the country outside the 'collection areas' of large-scale KCC plants.[4]
(4) Economies of scale might give larger-scale milk processing a significant cost advantage over smaller-scale plants. If this techno-economic obstacle is surmountable then the potential for decentralised development of the dairy industry would become clearer.[5]
(5) Small-scale dairy development involves linked technology choices for (a) product characteristics and (b) the method of production, and so is of broader interest in the context of case studies in technology performance and selection.[6]
(6) The ability of the management of small-scale plants in rural Kenya to master the technology of modern agro-food processing on a regular and systematic basis has been largely unproven, so that this further area of ignorance could also be clarified by the development of pilot plants.

This paper outlines the background to dairy processing in Western Kenya, considers the issues of technology choice and scale economies, and then summarises some of the conclusions of the discussion.

Background to Dairy Processing in Western Kenya

Recent development of the dairy industry in the Bungoma District of the Western Province of Kenya has been considerably assisted by a Finnish government aid programme. The programme has been the joint responsibility of the Kenyan Ministry of Agriculture/Livestock Development and FINNAGRO (the Finnish agricultural/agro-industrial agency undertaking this aspect of FINNIDA's aid programme). The Rural Dairy Development Project started in 1979 and has included baseline survey work, tick control, artificial insemination, pasture and fodder development, manpower development, equipment supply (for example, forage choppers and windmills/water pumps) and supply of milk coolers and processing equipment.[7] Finnish technical assistance personnel have also been provided. Part of the entire development programme has consisted of the establishment of a 'small-scale' dairy processing plant in Bungoma town involving about £1,000,000 investment and using the UHT (ultra high temperature) treatment technology.

Through upgrading, and increased size, of the dairy stock, disease control and improved feeding milk production yields and aggregate production have increased. This gives potential for higher on-farm milk consumption, but a limit is soon reached above which off-farm sales become inevitable so that export from 'milk-surplus' producing areas to 'milk-deficit' consuming areas becomes essential. It is at this stage that agro-industrial food processing and handling technology become very relevant.

Milk deteriorates rapidly in the high ambient temperatures of Western Kenya. Bacteria development is positively related to both temperature and time, so that within about seven hours of the raw milk leaving the cow it becomes unacceptable for further processing owing to high bacteria levels and other changes. Agitation (in the process of transportation) accelerates changes in the nature of the milk, so that poor roads which increase agitation and decrease speed present a major problem for dairy development (poor roads also increase costs through rapid deterioration of vehicles). The method conventionally used to overcome these problems is to have sufficient cooled storage capacity within easy reach of the farmer so that deterioration is significantly slowed very soon after milk is extracted from the cow.[8] In developed countries (and in the more developed parts of the dairy industry in less developed countries) this cooled milk is collected regularly by insulated road tankers (the tanks of which are baffled to reduce agitation).

The simplest processing option is simply to re-cool and sell the raw milk with or without packing, but with laboratory testing in order to check that it is fit for human consumption. Because the raw milk has not undergone any processing the bacteria which have developed from the time the milk left the cow are still present, and the 'shelf-life' is extremely limited. This is the option that was proposed in the original Bungoma project design published in the 1984 District Plan.[9] The use of milk coolers and refrigerators in the chain which extends from farmer to consumer can prolong the life of raw milk to a maximum of about two days.

The concern of this paper is primarily with two dairy products – pasteurised milk and ultra-heat-treated (UHT) milk. In both cases the raw milk is heated to a pre-determined temperature for a controlled period and is then re-cooled rapidly before packing.[10] As its name implies, UHT milk is heated to a higher temperature than pasteurised milk, but both techniques have the effect of destroying bacteria accumulated in the milk, and therefore of prolonging its 'shelf-life'. Pasteurised milk has a life span of between two and seven days with refrigeration, depending on the initial raw milk quality, but with UHT milk the 'shelf-life' may extend to two weeks to six months without refrigeration, depending on the method of packing employed. For lower income tropical countries, and particularly lower income groups, with low levels of domestic refrigerator ownership (and perhaps suffering from interrupted electricity supply) the extended life of UHT milk is an extremely attractive attribute.[11]

These product characteristics are extremely important in affecting the

TABLE 1

SIGNIFICANCE OF MILK DETERIORATION AND OF THE TECHNOLOGY LEVEL
IN THE KENYAN MARKET STRUCTURE

Stage in Milk Supply Chain	Raw Milk	Pasteurised Milk	UHT Milk	Powdered Milk
Farm	Critical/ Low	Critical/ Low	Critical/ Low	Critical/ Low
Raw Material Supply	Critical/ Low	Critical/ Low-Medium	Critical/ Low-Medium	Critical/ Low-Medium
Processing	Critical/ Low	Critical/ Medium-High	Critical/ High	Critical/ High
Wholesale	Critical/ Low	Significant/ Low-Medium	Not Critical/ Low	Not Critical/ Low
Retail	Critical/ Low	Significant/ Low-Medium	Not Critical/ Low	Not Critical/ Low
Consumer	Critical/ Low	Significant/ Low-Medium	Significant/ Low-Medium	Not Critical/ Low

nature of the market structure for dairy products. The limited 'shelf-life' of raw milk means that the raw material supply link from farmer to processor is of critical importance. However, the long life of UHT milk makes the link between processor and consumer much less critical with respect to time. The collection/delivery links between farmer and processor are therefore restricted geographically by the deterioration profile of the raw milk, while locational restrictions on the link between processor and consumer hardly exist at all (UHT milk can be traded internationally to some extent). By comparison, the geographical restrictions on the distribution of milk powder are even less than those on UHT milk, and may be limited more by the availability of potable water needed for reconstitution into liquid milk than by any other factor. The 'shelf-life' of milk powder is, of course, much longer than that for any form of liquid milk.

Table 1 attempts to place the issue of product characteristics in the context of the market structure:

Choice of Technology and Scale

The feasibility study for the two Kenyan small-scale dairy plants contains negligible consideration of the technological alternatives, particularly with respect to scale and product characteristics. The issues to be resolved are largely concerned with the selection of an appropriate (a) raw material, (b) product, (c) production technology, (d) scale of production, and (e) market structure.

Processing technology and scale of production can usefully be separated in the discussion which follows.

Processing Technology and Product Choice

The Kenya Cooperative Creameries have used the UHT technology for several years. There must, by now, be a body of labour familiar with the processing tech-nology. The precision nature of the heating and cooling of the milk (HTST – high temperature short time) involves sophisticated electronic control systems which require regular inspection and maintenance. This creates a demand for highly skilled Kenyan personnel although an experi-mental plant, such as that in Bungoma, can expect to have expatriate personnel and/or Kenyan 'headquarters' staff available for a period. Wider adoption of the technology would depend on a broader engineer-ing knowledge and experience amongst Kenyan operators. The training element of the Dairy Development Project is obviously addressed to this question. However, the manpower aspect of 'technology transfer' has to be recognised and tackled in this context.

The alternative technologies associated with raw milk cooling and packing, and of pasteurised milk production, involve technologies which are not as sophisticated as the modern HTST method. However, the production technology for milk powder would be as complex as that for UHT milk.

The packing machinery specified is designed to put milk into dated polyethylene sachets each of 500cl capacity (a 200cl pack can be produced as an alternative by adjusting the equipment). The characteristics of the sterile packing system give the 'Bungoma' milk a two-week shelf-life without refrigeration.

However, there is no great advantage to having a special packing technology if the milk is actually consumed within one or two days of processing, as is the case for the Bungoma plant. In a recent article Coughlin compares the costs of alternative milk packaging systems. He makes the point that: 'Much unpasteurised milk is also sold between rural neighbours and by vendors who ladle it from buckets. For instance, in Meru a quarter of the milk delivered to the milk societies is eventually sold this way since they earn more by selling raw milk directly to customers than to the dairy.'[12]

If the Bungoma project had been based on the cooling and packing of about 1,000 to 1,500 litres of raw milk per day as had originally been intended the packing technology issue would not have been exceptionally significant.[13] Consumers and retailers might use their own containers to collect milk, or the plant could itself supply some containers. When the project was altered to the UHT option packing became an issue. Coughlin particularly argues against the Tetrapak (as used by the KCC), and in favour of polyethylene sachets (as used in Bungoma – although he does not mention the Bungoma project) on the grounds of both cost (expressed in Kenya Shs) and foreign exchange content. The Tetrapak is a substantial 'waxed carton' type container costing about Shs0–74 per litre of packed milk in 1984/85 (13 per cent of the retail price of milk), as against about Shs0–45 for polyethylene sachets. The direct foreign exchange costs were Shs0–23 for Tetrapak and Shs0–13 per litre for sachets.[14]

The additional cost of the Tetrapak system gives a three-month unrefrigerated shelf life, as compared with the two-week shelf life for the sachets used in the Bungoma project. As long as this difference is clearly appreciated by traders and consumers it is unlikely that the shorter shelf life of the sachets really represents any great handicap. Because of the storage characteristics of the UHT milk, wholesalers and retailers do not need refrigeration equipment so long as the milk is sold to consumers before two weeks have elapsed from the time of processing. The date of processing is recorded on each sachet of milk.

In 1986 a detailed survey of dairy development in Meru was published by the Institute of Development Studies, University of Helsinki, prepared by a team of two Finns and two Kenyans.[15] The study contains a wealth of information about consumption habits and livestock development, and an interesting chapter on the School Milk Scheme which the Bungoma project had been aspiring to supply in their market area: 'The National School Milk Programme aims at distributing milk free to the pupils of all the primary schools in the country. At present, every child is to get two decilitres of milk twice a week during the school terms, and the deliveries are organised through Kenya Cooperative Creameries.'[16]

A further quotation is appropriate: 'The storage time of UHT milk depended on the amount of milk brought. Milk was normally stored for a month or more, sometimes a whole term (3 months). A number of packets had been spoiled during storage. During the first month the quality had been good, but towards the third month about 50 per cent of the packets had been spoiled ... UHT milk was stored in a special milk store room (21 schools), in a general school store (19 schools), or in the staff room or headmaster's office (3 schools).'[17]

The School Milk Scheme is perhaps the major 'bulk' order that the Bungoma plant could aspire to supply. As of 1987 the KCC had a monopoly supply position, and the Ministry of Education had not permitted any of the smaller dairies to join the scheme. Significantly, the price paid per litre of milk supplied to the School Milk Scheme was higher than that for general milk sales, and this was recognised in the FIN-NAGRO feasibility study for the project. The quotation from the Meru study makes it clear that bulk delivery of milk to schools in 5 or 50 litre churns is unlikely to be practicable, particularly for schools in the more rural areas, so that UHT milk in a packed form is the more likely option given the general nutritional benefits gained from the Scheme.[18] Equally, the longer shelf life of the KCC Tetrapaks is a significant advantage in the circumstances described by the University of Helsinki study.

Scale of Production

There is some evidence, admittedly not from the dairy industry, that the nature of scale economies, arising from industrial equipment and associated investment, is such that smaller-scale plants using sophisticated technologies have a substantial cost disadvantage compared with larger-scale plants using the same technology type.[19] This is not to say that industries which almost universally have small-scale plants, and which employ sophisticated equipment, will be comparatively inefficient. Rather, the argument is that where large and small plants using the same technology type coexist in the same industry, then the smaller plants will often have a serious cost disadvantage due to the incidence of high initial investment costs per unit of output.

Most of the disadvantage arising from the smaller scale is based on the '0.6 rule'. Investment costs tend to increase significantly less than proportionately with capacity as plant size is increased. Because the relationship is likely to be curvilinear (see Figure 1) the smallest scales of operation suffer the greatest proportionate cost disadvantage. Above a certain scale of production there tends to be no 'intrinsic' advantage from increasing the scale still further – that is, scale economies (on account of this '0.6' factor) are eventually exhausted.

There are other intrinsic 'scale' factors disadvantageous to small-scale processing plants using the same technology type as their larger-scale counterparts. Perhaps the most important relates to the labour force. Whether a particular item of equipment is large or small it tends to need the same number of people, with similar skill levels, to operate it. This means that the number of equipment operators required to run a 1,000 litres per hour plant is likely to be approximately the same as for a 10,000 litres per hour plant if increased capacity is achieved through larger-sized pieces of machinery. Costs per unit of throughput for processing labour would therefore be likely to fall directly proportionately to the increase in

FIGURE 1

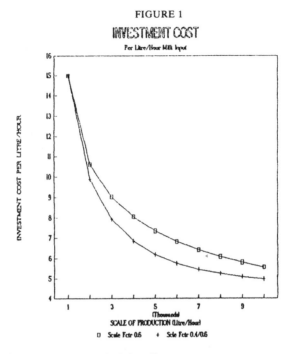

Source: Based on illustrative calculations.[20]

the capacity of the equipment. Of course, if increased scale is achieved solely and simply by replication of smaller units of equipment then this particular advantage of larger scale operation will not apply. The scale benefits arise from the increased size of single pieces of equipment within the same technology type.

The behaviour of labour costs is often similar for non-processing manpower since, in many cases, the number of skilled personnel required to run a larger-scale plant is proportionately smaller than that required to run a smaller-scale plant. This applies, for example, to the supervision of production, administration and marketing. While some savings might be possible in the organisation of smaller-scale plants (with doubling up on management functions through reduced specialisation of tasks or through using less highly trained manpower), in general a reduction in the quality of the manpower, or an increase in the number of tasks undertaken by particular individuals (reducing specialisation), might be expected to reduce the effectiveness of the management, and to reduce the efficiency of production at the smaller scale, thus increasing unit production costs.

It should be emphasised that these remarks relate to different scales of plant using essentially the same technology type. In many industries small-scale and large-scale plants use somewhat different technologies so that, for example, small-scale plants cooling and packing raw milk might coexist with very large-scale plants producing UHT milk and powdered milk.

We might expect, therefore, that unit costs for packed UHT milk are higher for the Bungoma UHT plant than for the KCC. Because the small-scale plant, as a price-taker, has to operate within the same price regime as the large-scale KCC plants it is not possible for this cost disadvantage to be offset through acquisition of raw milk from member and non-member farmers at a lower price than is paid by the KCC, or to sell its processed milk at a higher price in the market (either to wholesalers or to retailers – milk is not sold directly to the consumer). The Bungoma plant would therefore have to look for lower distribution, overhead and other costs in order to maintain competitiveness with the KCC.

Conclusions

One of the principal objectives of the development of small-scale dairy processing in Kenya has been the creation of additional employment and incomes in rural and small-town areas. The intention was that the milk sold to smaller-scale dairy processors should not be diverted from the large-scale plants of the Kenya Cooperative Creameries (KCC) so that the two would complement each other.

A number of issues arise from the discussion in this paper:

(1) In terms of small-scale rural industrialisation the UHT dairy plant in Bungoma cannot be regarded as 'small-scale'. Neither the level of employment (in excess of 50 persons) nor the capital investment employed (in the order of £1,000,000 sterling) can be regarded as 'small'. The original project design, involving cooling and packing of raw milk for sale, would probably be regarded as small-scale in terms of both employment and capital employed.

(2) However, the Bungoma plant is certainly not 'large-scale'. The KCC plants have a capacity in the order of 250,000 litres per day compared with the Bungoma capacity of 8,000 to 12,000 litres per day. Perhaps 'medium-scale' would be an appropriate description.

(3) It is possible that the farm income and nutritional objectives of the project could have been achieved without resorting to the UHT technology. Limiting the product type to raw (unpasteurised) milk would have reduced the technological impact, but would perhaps not signifi-

cantly have reduced the economic impact, of the project. This issue can be linked to two questions:

(a) To what extent is UHT milk the appropriate product choice in the Kenyan rural processing context for any plants other than the large-scale KCC dairies? The development of powdered milk as a 'longer shelf life' alternative by the KCC reinforces the doubts about adoption of the UHT process at the small/medium scale.

(b) To what extent should more complex ('high') technologies be adopted in the 'Bungoma' context – i.e. more remote, smaller-scale, projects? Effective operation and maintenance of such technologies depends upon the existence of an appropriate local technological environment, and particularly the availability of personnel who have the appropriate skills for the UHT technology. While overseas technical assistance can provide valuable support for a limited period, it is not clear that effectiveness can be sustained in the longer term. Thus, the use of highly skilled personnel in small-scale plants might be seen as an economically inappropriate utilisation of those skills.

(4) The UHT plant which was installed in Bungoma was made available 'free of charge' under a generous international aid programme. The financial viability of the project was thus virtually assured. However, the longer-term economic viability is much more open to question. It is unlikely that *all* similar plants would in future have the major elements of capital equipment donated, particularly if they were to be established in the private sector (rather than in the 'public' co-operative sector). Similar experiences have occurred with the provision of 'small-scale', high technology, sugar processing plants in the recent past, where the equipment was essentially donated.[21] Such pilot plants can yield valuable experience, but the longer term economic and financial viability of the technology are open to question.

(5) Competition with the large-scale and nationwide Kenya Cooperative Creameries means that while the farmers who supply raw milk to Bungoma could not, in general, alternatively send their milk to the KCC plants, it is quite easy for the KCC to provide UHT milk to the entire market aimed at by the Bungoma plant. The issue of how the smaller-scale plant 'breaks-in' to the pre-existing market structure is therefore a major one.

These five issues should be sufficient to illustrate the wide range of ramifications arising from the attempt to introduce 'high technology' into the smaller-scale end of the spectrum of dairy processing in Kenya.

The broad conclusion might well be that there is really no place in the immediate future for the type of smaller-scale UHT plant which was the subject of the 'pilot project' discussed in this paper. The combination of scale and technology falls uneasily between the large-scale 'high technology' dairy plants (which, *inter alia*, have a wide range of product types – including UHT milk and milk powder) and smaller-scale, relatively simple, plants which simply cool and pack raw milk.

The implications of this type of conclusion might well include a more limited view for enhancement of the productivity of smaller-scale processing activities. In turn this would also affect the view of the potential for a significant technological leap at the smaller-scale end of the processing spectrum – and, perhaps, lead to the conclusion that dualistic development (both economic and technological) is likely to be with us for some considerable time to come – particularly given the population dynamics of a country such as Kenya.[22]

NOTES

This paper is a shortened, revised and developed version of 'Developing Small-Scale Agro-Processing in Kenya: Lessons from a Post-Implementation Review', No. 10 in the Development and Project Planning Centre's new Discussion Paper series. Acknowledgements to the British Council, Kenya Institute of Administration, FINNIDA and to individuals in the discussion paper also apply to this conference paper.

1. See for example: Republic of Kenya, *Sessional Paper No.4 of 1981 on National Food Policy*, paras 4.41 and 4.43, pp.35–6, Government Printer, Nairobi, 1981; Republic of Kenya, *National Livestock Development Policy*, para 77, p.38, Government Printer, Nairobi, 1980; Republic of Kenya, *Development Plan 1983–1988*, para 6.202(ix) p.181, Government Printer, Nairobi, 1983; Republic of Kenya, *Development Plan 1989–93*, Government Printer, Nairobi, 1989.
2. Republic of Kenya, *District Focus for Rural Development* (as revised – March 1987), Government Printer, Nairobi, 1987. See also Sessional paper No.1 of 1986, *Economic Management for Renewed Growth*, Government Printer, Nairobi, 1986, para 1.8, p.3 and para 4.1, p.41, and I. Livingstone, *Rural Development, Employment and Incomes in Kenya*, Gower for the International Labour Organisation, Aldershot, 1986, particularly Chs. 2 and 6.
3. Republic of Kenya, *Development Plan 1983–1988*, para 1.68 p.33 and paras 6.218 to 6.221 pp.186–7 on 'Food and Nutrition', op. cit.; Republic of Kenya, *Sessional Paper No. 4 of 1981 ...*, para 4.75, p.43, op. cit.; P. Coughlin, 'Development Policy and Inappropriate Product Technology', *Eastern Africa Economic Review* (New Series), Vol.4, No.1 (June 1988), 28: A. Karinpää, R. Launonen, L. Marangu and S. Minae, *Rural Dairy Development in Meru*, University of Helsinki, Institute of Development Studies, 1985 (Report 8/1985 B), see Ch.6 – The School Milk Scheme; A. Singh, 'The Basic Needs Approach to Development vs the New International Economic Order', *World Development*, Vol.7, No.6 (June 1979).
4. Republic of Kenya, *Sessional Paper No.1 of 1986*, op. cit. paras 1.5 and 1.6, p.2 and para 2.11, p.10.

5. M. A. Tribe, 'Scale Considerations in Sugar Production Planning', in R. Kaplinsky *et al.* (eds.), *Cane Sugar: The Small-scale Processing Option* (London: Intermediate Technology Publications, 1989).
6. J. James and F. Stewart, 'New Products: A Discussion of the Welfare Effects of the Introduction of New Products in Developing Countries', in F. Stewart and J. James (eds.), *The Economics of New Technology in Developing Countries*, Frances Pinter, London, 1982 – reprinted from *Oxford Economic Papers*, Vol. 33, No.1 (1981).
7. Kenya Institute of Administration – Project Development and Management Course No.30/86, 'An Evaluation of Milk Production, Processing and Marketing in Bungoma District', Nov. 1986, mimeo., and T. Lundström, 'Profitability of Smallholder Dairy Farms in Bungoma District, Kenya', Finnagro Oy, Helsinki, 1988, mimeo.
8. See the Technical Appendix to M.A. Tribe, *Developing Small-scale Agro-Processing in Kenya: Lessons from a Post-Implementation Review*, Development and Project Planning Centre, University of Bradford, Discussion Paper No.10 (New Series), 1989.
9. Ibid.
10. The issue of product characteristics is discussed in the references cited in notes 3 and 6.
11. See the Technical Appendix in M.A. Tribe, Discussion Paper, op. cit.
12. P. Coughlin, op. cit., p.28.
13. Republic of Kenya, *Bungoma District Development Plan 1984–1988*, Ministry of Finance and Planning, Nairobi, 1984, p.76 and Annex pp.23–4.
14. Coughlin, op. cit., p.29.
15. A. Karinpää *et al.*, op. cit.
16. Ibid., p.107.
17. Ibid., p.108.
18. Ibid., pp.109–10.
19. M.A. Tribe, 'Scale Considerations . . .', op. cit; M.A. Tribe and R.L.W. Alpine, 'Scale Economies and the "0.6 Rule"', *Engineering Costs and Production Economics*, No.10 (1986), 271–78, and M.A. Tribe and R.L.W. Alpine, 'Sources of Scale Economies: Sugar Production in Less Developed Countries', *Oxford Bulletin of Economics and Statistics*, Vol. 49 No.2 (May 1987), 209–26.
20. In Figure 1 the higher of the two curves shows a constant 'scale factor' of 0.6, and the lower curve a scale factor varying from 0.4 at the smallest scales to 0.8 at the largest – giving a stronger cost advantage to medium and larger scale plants in the lower curve case. For the 'vacuum pan' technology in the cane sugar industry the lower curve appears to approximate to reality, and a similar situation *might* apply to the dairy industry – see Tribe and Alpine, 1986, op. cit.
21. A. James, 'New Cane Extraction Technologies for Small Scale Factories', in R. Kaplinsky *et al.* (eds.), op. cit., pp.61–3 – see especially footnote on p.63.
22. Republic of Kenya, *Development Plan 1989–1993*, Government Printer, Nairobi, 1989 – see especially 'Contribution to Rural–Urban Development Strategy', pp.112–13 and 'Employment Creation Strategies', pp.196–202.

The Semiconductor Industry: Its Contribution to Third World Development

ALISTAIR YOUNG

The contribution of high technology industries to the development of LDCs has frequently been questioned, particularly by proponents of the 'New International Division of Labour' (NIDL) hypothesis. Critics have focused on the role of the transnationals which have commonly served as the vehicle for technology transfer; it has been alleged that spinoffs to the local economy have been limited or non-existent, while the workforce has had to endure unhealthy and exploitative work conditions in the new industries. This paper reviews these criticisms in the particular context of the East Asian semiconductor industry. It is argued that the most recent evidence from the major locations of the industry supports the view that its contribution to local economic development has been on balance beneficial, both in countries where the industry continues to be dominated by transnationals and in others where a substantial indigenous capacity has emerged.

Introduction

Can high technology industries help to make poor regions prosperous? Government agencies seem to think so, whether they have the task of promoting industrial renewal in the stagnating regions of the long-industrialised countries, or of building up a modern industrial base in a Third World economy. Academic commentators, however, have not always been so sure.

Their reservations about the abilities of high technology industries to create prosperity are frequently concerned with the question: who is to

Professor A. Young is in the Department of Economics and Management, Paisley College, Paisley, Scotland.

make this high technology available, and on what terms? New technology cannot be bought in as an isolated input; it generally comes (often through the medium of transnational corporations) as part of a package deal which requires the recipient region to become progressively incorporated, to a greater or lesser degree, within the international economy. This process of incorporation may be considered an essential part of modernisation and economic development; or, more pessimistically, it may be seen as a means whereby the poorer countries are allocated to a permanently subordinate and dependent role within the world economic system, not for their own benefit but for the benefit of the powerful interests which dominate that system.

Perhaps the clearest and gloomiest account of the implications for Third World countries of incorporation within the global economic system is given by the 'New International Division of Labour' (NIDL) hypothesis. The fullest statement of this is to be found in Frobel, Heinrichs and Kreye [*1980*].

These writers argue that the 'Taylorist' fragmentation of production, on the one hand, and the development of modern communications on the other, have together allowed certain labour-intensive stages of production to be established in low-cost labour locations. But the host countries do not benefit from this phenomenon. Because they merely process semi-finished imports for re-export, there is little scope for backward or forward linkage. There is equally little technology transfer; control of technology remains firmly in the corporate headquarters in the rich countries. The workers, for their part, suffer 'super-exploitation'; they are worked to exhaustion in hazardous conditions, and then replaced from a rural 'reserve army'. The presence of this pool of unemployed or underemployed labour inhibits attempts of employed workers to organise to demand better conditions. Further to ensure a docile workforce, firms often recruit chiefly among young women.

The role of host-country governments, according to the NIDL hypothesis, is to provide an environment which will attract the transnationals. They do this in two ways: by restrictive labour legislation (or, if necessary, armed force); and through institutions such as 'export processing zones' (EPZs), in which firms are given tax holidays and assured of minimal restrictions on repatriation of profits. This ensures that little of the surplus produced in the 'world market factories' gets ploughed back into the host economy. The host governments collaborate in this process either because they are corrupt or, perhaps, because they are unfamiliar with the NIDL hypothesis.

The NIDL hypothesis, as presented by Frobel and his colleagues, has been much criticised [*Jenkins, 1983; Henderson, 1986, 1989; Hill, 1987;*

Wield and Rhodes, 1988]; the critics have argued, *inter alia*, that it gives sufficient weight neither to the diversity of motive of the transnationals, nor to the diversity of experience of the host countries, nor to the scope for independent action by their governments. The remainder of this paper seeks to review the current state of the debate as it relates to one industry in particular – the manufacture of semiconductors – and one group of Third World countries which has encouraged the establishment of this industry. By so doing, it is hoped to provide at least a partial answer to the question posed at the beginning.

The Semiconductor Industry in East Asian Third World Countries

The semiconductor industry is of particular interest in the present context. It is difficult to overestimate the importance of the integrated circuit, which has become the basis of the key 'enabling technology' of the late twentieth century. As Ernst notes, the semiconductor industry is one 'which is widely considered to be of crucial importance for future structural transformation and international competitiveness' [*Ernst, 1987: 38*]. Thus it may be argued that it is an industry against which theories of world economic restructuring might reasonably be tested, if such theories are to have any claim to generality; and, indeed, it does feature as an industry to which frequent reference is made in the study by Frobel *et al.*

It is also an industry which is particularly characterised by the global fragmentation of production. Accounts of the process of semiconductor production generally identify five main stages: wafer manufacture, circuit design, wafer processing, assembly and testing. Of these, all except assembly are capital-intensive. In the case of assembly, there is a choice of technique; and while Japanese manufacturers have opted for automated methods, American companies have tended to use manual assembly, often in Third World locations.

The process of establishing offshore semiconductor plants in Third World Asian countries began with Fairchild's assembly plant in Hong Kong in 1962; other American companies followed suit, with the main locations for start-ups being Hong Kong, Singapore and South Korea in the 1960s, and Taiwan, Malaysia, Thailand, the Philippines and Indonesia in the 1970s. Even by the early 1980s the great majority of these plants were confined to the labour-intensive assembly operations. Research and development, design, wafer manufacturing and wafer processing continued to be located in developed countries. There was, however, provision for final testing in several Asian subsidiaries.

It is true that the Asian semiconductor industry did not match the NIDL hypothesis in all respects; Hong Kong and Singapore are not examples of 'rural labour-surplus' economies. Yet the labour-cost reductions from choosing a Third World location were undoubtedly substantial. As a study by Scott [1987] shows, in the late 1960s the level of hourly compensation of electronics production workers in Hong Kong and South Korea was one-tenth of the figure for the United States; the ratio in Singapore and Taiwan was even less. Other aspects of the industry seemed to provide further support for the 'exploitation' hypothesis. The newly-established assembly plants made intensive use of women workers; these workers were 'young, single, aged between 16–24, with basic primary education, and holding their first job', and generally drawn from the rural areas [Salih et al. 1988: 393]. The conditions which these workers faced included long hours in an unhealthy environment; estimates quoted in Lin [1987: 121] suggest that in the Export Processing Zones in Malaysia a quarter of workers had a work-week of over 48 hours, while in the Philippines a similar proportion had hours in excess of 60. Other problems noted by the same author include the hazards for reproductive health arising from radiation and gas leakages, and deterioration of eyesight through intensive work with microscopes.

In the study by Frobel et al., much emphasis was placed upon the restrictions on worker organisation in the Third World countries which had succeeded in attracting transnational manufacturing plants. In the case of Malaysia, for example, strikes were effectively outlawed under the 1967 Industrial Relations Act, while in South Korea they were banned under the emergency measures of 1971, subject to a penalty of up to seven years in prison [Frobel et al.: 360–64].

While it is clearly possible to find evidence from the offshore semiconductor industry which conforms to the picture painted by the 'strong' version of the NIDL theory, there is also a growing body of work which suggests that the industry is developing in ways which are not so easily reconciled with the theory; and which perhaps give more grounds for optimism regarding its contribution to the prospects for economic progress of the host countries.

Recent Developments

One of the most marked features of the economies which have served for the longest time as a location for the offshore semiconductor industry is the strong upward trend in the rewards to labour. The study by Scott,

TABLE 1
RATIO OF ASIAN TO US HOURLY COMPENSATION: PRODUCTION WORKERS
IN ELECTRONICS (US = 100)

	1969	1985
Hong Kong	10	16
Republic of Korea	10	14
Taiwan	8	16
Singapore	9	19
Indonesia	n.a.	4
Malaysia	n.a.	10
Philippines	n.a.	8
Thailand	n.a.	5

Source: From Scott (1987), p.145.

mentioned earlier, shows a narrowing of the gap between US manufacturing wages and those in Hong Kong, Taiwan, Singapore and the
Republic of Korea. The relevant data are given in Table 1.

At first sight, these figures suggest that the workers in the newly
industrialising countries have benefited substantially from their participation in the restructuring of the world economy. However, such figures
have also been given a more pessimistic interpretation by some commentators. If the comparative labour cost advantages of the offshore locations
are being eroded in this way, for how long will they be able to continue to
attract or retain the transnational companies? Will the TNCs seek out
other 'unspoilt' locations, or even attempt to substitute capital-intensive
for labour-intensive methods of production?

The scope which existed in the mid-1980s for the first of these strategies
is also illustrated in Table 1. It is clear from the figures that the relative
latecomers to the offshore business – Malaysia, Thailand, Indonesia and
the Philippines – have a significantly greater labour cost advantage than
the more established locations.

The second course of action, to substitute automated for labour-
intensive techniques, further complicates the picture. In the early
1980s, there was some discussion of the possibility that offshore semiconductor assembly might be progressively 'repatriated', in part as a
natural response to rising wages, and in part as a consequence of the
technological and organisational developments within the industry itself.
Automated assembly made sense in terms of ensuring more rigorous
quality control; while the adoption of 'Just-in-Time' manufacturing,
implying very close relationships between suppliers and users of inputs,

TABLE 2
MANUFACTURING COST PER INTEGRATED CIRCUIT DEVICE, HONG KONG
AND THE UNITED STATES, 1982 ($US)

	Hong Kong	US
Manual	0.0248	0.0753
Semiautomatic	0.0183	0.0293
Automatic	0.0163	0.0178

Source: Electronics, 11 August 1982 (cited in UNCTC, 1983, p. 1).

might be facilitated by greater geographical proximity, which it was
assumed would imply centralisation in the 'developed' rather than the
'offshore' locations. With automation, the cost advantage of an offshore
location is much reduced, as Table 2 shows.

Yet despite these factors tending to encourage repatriation, no marked
tendency in that direction has become apparent. Instead, what seems to
be happening is a relocation of specialisation within the East Asian region
[Ernst, 1985; Henderson, 1986]. The longer-established, higher cost
centres – Hong Kong and Singapore – have moved into small-batch, more
technologically sophisticated chip production, leaving the more stan-
dardised chips to Malaysia, the Philippines, Indonesia and Thailand.

The Development of Indigenous High Technology Enterprises

The assessment of the role of high technology companies in Third World
countries is further complicated by the appearance of indigenous enterprises in the high technology sectors. This phenomenon has been particularly characteristic of the Korean economy in recent years.

Like Japan, but by contrast with its other East Asian competitors,
Korea has tended to discourage foreign direct investment. The electronics sector, to be sure, has been a partial exception to this general
principle; by the mid-1970s, 75 per cent of electronics exports were produced by foreign firms. Thereafter, however, the growth of indigenous
firms reduced this proportion [van Liemt, 1988]. Among the Korean
firms which participated in this expansion were the chaebol, or large
conglomerates. During the slump in the semiconductor market in the
mid-1980s, the electronics subsidiaries of two of these conglomerates,
Samsung Electronics and Hyundai Electronics, made heavy investments
in semiconductor fabrication at a time when US and Japanese chipmakers
were retrenching [Electronics, 28 April 1988: 21]. Korea also has the

largest contract assembly plant in the world, Anam Industrial Co. Ltd; this firm too made large-scale investments during the recession.

But will this bid to achieve a strong and independent market presence pay off? Ernst is pessimistic, judging that 'the chances for successful latecomer entry into the worldwide IC commodity markets have become quite low' [1987: 52]. The most recent indicators seem to lend some support to this view [Electronics, January 1990: 91–2]. After a recovery in the late 1980s, the world semiconductor market is now once again over-supplied with low-cost memory chips, reflecting a stagnating demand for personal computers. This has pushed Korean prices to the point where manufacturers are in danger of being accused of dumping in the US market.

Long-Run Prospects

In the long run, however, the difficulties for indigenous enterprises of securing a foothold in the commodity end of the market may simply encourage them to seek to exploit the more specialised, design-intensive end. Recently, there has been a movement by Korean producers into the manufacture of 'application-specific' ICs (ASICs). This has also been the strategy of Taiwanese design houses, only a minority of which are foreign subsidiaries. In the Taiwanese case in particular, this strategy has been assisted by (and has in turn assisted) the development of strong linkages between the semiconductor sector and the information technology equipment sector.

Of course, even if the strategy pursued by the more developed NICs – Korea and Taiwan, and perhaps also Singapore and Hong Kong – succeeds in building up a balanced electronics complex based on leading-edge semiconductor technology, it does not follow that the other East Asian countries which currently host semiconductor assembly will be able to follow suit. In fact, however, some of these countries seem closer to achieving this than others. Salih et al. note that Indonesia has so far proved an unattractive location to TNCs in this industry because of 'rampant corruption and bureaucracy, poor infrastructure and limited skilled manpower', while Thailand faces the disadvantage of not having an English-speaking workforce and the Philippines suffers from political instability [1988: 377–9]. The industry in Malaysia, however, and in particular the cluster of semiconductor firms in the Penang EPZ, seems to be well established. The most recent evidence is that a process of backward linkage is beginning to emerge [ibid.: 383 ff.]. Although Hewlett Packard rejected Malaysia as a location for wafer fabrication in 1986, in favour of Singapore, National Semiconductor decided to set up a wafer

fabrication plant in Penang in 1987. The same company has in the past
established other 'upstream' subsidiaries in Malaysia to supply both its
own assembly plant and other semiconductor firms.

Recent Evidence on Work Conditions

Studies of the South-East Asian semiconductor plants by Lin [*1987*] and
Salih *et al.* [*1988*] help to bring out the advantages as well as the drawbacks of the industry from the perspective of the local workforce.

Lin's study is based on a sample of women workers in Penang and
Singapore. Her findings indicate that by the mid-1980s the age of the
workers had increased, as compared with the ages reported in earlier
studies; rather than the workforce being confined to the late teens and
early twenties, the average age in her sample was 24 years in Penang,
while in Singapore it was 27.5. Again by contrast with previous received
wisdom, marriage did not automatically lead to departure from the
labour force; in Singapore, 46.5 per cent of the workers were, or had been
married, the corresponding figure for Penang being 24.5 per cent. In each
case, over three-quarters of the married women workers had children.
Lin's findings also challenge the notion that workers leave the plant when
'worn out'; most workers who leave do so within the first year (a high
first-year turnover rate is, of course, a feature common to plants in
developed industrial countries as well). She notes that while occupational
health problems continue to be associated with the production processes
involved, improvements in safety have been made within the plants
studied, though this may partly be a consequence of increasing automation.

Although the work is still 'tedious and demanding', involving long and
frequently unsocial hours (which limit the opportunities for further training through night-school), and subject to the threat of layoffs through
fluctuations in the market for semiconductors, the picture which emerges
from Lin's account does not match at all closely with the 'superexploita-
tion' hypothesis. Most of the workers in the sample (85 per cent) considered that their lives had improved as a result of employment in the
semiconductor industry. Lin argues that a permanent, and increasingly
assertive, working class is emerging amongst the women workers within
the industry; although this trend is perhaps more marked amongst
workers of Chinese rather than Malay ethnic origin.

The study by Salih *et al.* shows that, as a result of the progressive
introduction of automation within the Malaysian offshore industry, the
labour force is becoming rapidly more skilled, to the extent that some
Malaysian engineers are being sent to the American parent company

to train American workers in assembly and testing; others 'interface' between the design team in the corporate headquarters and the customers. On the downside, the development of a more highly skilled labour force is associated with a need for fewer direct production workers. There has also been a shrinkage of the core labour force, as workers are dismissed in recession but rehired, when needed, only on short-term contracts; a move towards 'employment flexibility' which, of course, has parallels in the advanced industrial countries.

Conclusion

From the evidence surveyed in this paper, it appears that the question posed at the beginning may be answered with a cautious affirmative, at least as far as the semiconductor industry is concerned. On balance, the Third World countries of East Asia which have become locations for this industry have benefited from its presence, and seem likely to go on doing so into the future, as subsidiaries of transnationals develop backward and forward linkage, and as indigenous producers seek out 'niche' markets.

Employees in the industry work long hours for very low pay, by Western standards; they suffer from work-related illnesses, and experience job insecurity brought about by the sharp fluctuations in demand which have characterised the semiconductor market. Yet their pay and conditions, although inadequate, have been improving; and an increasing proportion of the workforce are receiving training to a high level of skill. The workers themselves perceive the industry as enhancing the economic opportunities available to them.

On the whole, then, the enthusiasm of Third World governments for this particular high technology industry would seem to be justified. No doubt it offers mixed blessings; but they are blessings nevertheless.

REFERENCES

Dixon, C., Drakakis-Smith, D. and H. Watts (eds.), 1986, *Multinational Corporations and the Third World*, London: Croom Helm.
Ernst, D., 1985, 'Automation and the Worldwide Restructuring of the Electronics Industry: Strategic Implications for Developing Countries', *World Development*, 13 (3), pp.333–52.
Ernst, D., 1987, 'US–Japanese Competition and the Worldwide Restructuring of the Electronics Industry: A European View', in Henderson and Castells [*1987: 38–59*].
Frobel, F., Heinrichs, J. and O. Kreye, 1980, *The New International Division of Labour*, Cambridge: Cambridge University Press.
Henderson, J., 1986, 'The New International Division of Labour and American Semiconductor Production in Southeast Asia', in Dixon *et al.*, pp.91–117.

Henderson, J., 1989, *The Globalisation of High Technology Production*, London: Routledge.

Henderson, J. and M. Castells (eds.), 1987, *Global Restructuring and Territorial Development*, London: Sage.

Hill, R. C., 1987, 'Global Factory and Company Town: The Changing Division of Labour in the International Automobile Industry', in Henderson and Castells [*1987: 18–37*].

Jenkins, R., 1983, 'The "New International Division of Labour": A Survey of Positions', University of East Anglia School of Development Studies, Discussion Paper 151, Norwich.

Lin, V., 1987, 'Women Electronics Workers in Southeast Asia: Emergence of a Working Class', in Henderson and Castells [*1987: 112–35*].

van Liemt, G., 1988, *Bridging the Gap: Four Newly Industrialising Countries and The Changing International Division of Labour*, Geneva: ILO.

Salih, K., Young, M.L. and R. Rasiah, 1988, 'The Changing Face of the Electronics Industry in the Periphery: The Case of Malaysia', *International Journal of Urban and Regional Research*, 12(3), pp.375–403.

Scott, A.J., 1987, 'The Semiconductor Industry in South-East Asia: Organization, Location and the International Division of Labour', *Regional Studies*, 21 (2), pp.143–60.

United Nations Centre on Transnational Corporations, 1983, *Transnational Corporations in the International Semiconductor Industry*, New York.

Wield, D., and E. Rhodes, 1988, 'Divisions of Labour or Labour Divided?' in B. Crow, M. Thorp *et al.*, *Survival and Change in the Third World*, Oxford: Polity Press/Basil Blackwell.

Summary Papers

The Transfer of Medical Technology to Less Developed Countries

DAVID NEWLANDS

Most debates about the nature and appropriateness of the technology transferred from developed to less developed countries centre on the machinery used in the production of physical goods. However, while the transfer of methods of health care is partly embodied in physical goods, such as surgical operating equipment, human skills and attitudes are also, if not more, important. The analysis of medical technology thus points to features of technology transfer which have previously been neglected.

In developed countries, there is a pervasive but very specific, and historically recent, view of health and health care. 'Modern medicine' focuses on the cure of illness within an elaborate institutional framework dominated by the specialised expert and employing capital-intensive, hospital-based technology. For people in most less developed countries, access to this type of health care is very limited. Nevertheless, the aim of many governments has been to increase the availability of 'modern medicine'. Scarce resources have been concentrated in intensive forms of health care, such as large hospitals, which are mostly to be found in the cities.

As in so many other areas, transnational companies are one of the principal vehicles for the transfer of medical technology, particularly drugs and surgical equipment. However, the assumptions which doctors and health administrators make about the causes of ill health and the appropriate treatment of it are equally important. A crucial element in the transfer of ideas is the professional medical training received by doctors in less developed countries. Most doctors came from wealthy families and are educated, at very great expense, in medical schools in developed countries or local medical schools with similar curricula.

Some less developed countries have adopted policies intended to achieve a more equitable distribution of health services. Specifically, they have sought greater equality of access to health services, a decisive commitment to improve preventative measures, and the increased participation of communities in local health programmes. Such countries have not necessarily wholly rejected the priorities and values of 'modern medicine', but all exhibit a concern to maintain indigenous systems of health care.

David Newlands is in the Department of Economics, University of Aberdeen.

Application of Appropriate Technology to Small-Scale Hydroelectric Power

A.A. WILLIAMS, N.P.A. SMITH, S. MATHEMA

This paper is based on an article originally published in *Science Technology and Development*, Volume 7, Number 2 (August 1989) with some additional material about technology transfer to bring it up to date.

There are several possible ways to harness the energy of mountain rivers and streams, some of which rely on large civil engineering works and sophisticated equipment imported from industrialised countries. This paper describes work which has been done, mainly in Nepal, on much smaller-scale systems which maximise the use of local skills and materials.

Water power is being developed for generating electricity to provide lighting more cheaply than using kerosene by adapting modern technology and incorporating traditional knowledge of water mills and irrigation systems. In particular, the use of standard electric motors as generators coupled to locally manufactured turbines, has been found to provide a suitable electricity supply for remote village communities. However, variations in electrical load can cause damaging fluctuations in voltage. A simple load controller has been developed in the UK to improve this system, which could find application in many developing countries. With an induction generator and controller, it is economic to supply electricity 24 hours a day, and work is being done to develop suitable end-uses for the power – for example, slow cookers and crop dryers.

The possibility of using standard pump units as turbines is also being investigated. These may be suitable for countries where, unlike Nepal, water turbines are imported, but pumps, which can easily be adapted for the same purpose, are already manufactured in large numbers.

A. A. Williams and N. P. A. Smith are Researchers in the Department of Electrical and Electronic Engineering, Nottingham Polytechnic, UK, working in collaboration with Intermediate Technology Development Group, Rugby, UK. S. Mathema is an Electrical Engineer with Kathmandu Metal Industries, Kathmandu, Nepal.

Small-Scale Engine-Powered Mechanisation and Transport Devices for the Rural Sector of Developing Countries

PETER CROSSLEY

Agricultural Mechanisation

The draught requirements to provide single-tine cultivation in the dry season is stated to be around 4–5kN. Small-scale mechanisation devices must provide this pull efficiently. Many attempts have been made to produce a satisfactory small tractor, but the provision of suitable thrust performance from small tyres with low weight is difficult. Various three-wheeled and four-wheeled machines are described.

Alternative winch-based devices designed at Silsoe College are introduced. The Snail (2-wheeled) and Spider (4-wheeled) machines combine winching, for high draught operations, with direct traction for low draught and transport. Fuel consumption is far better than for an equivalanet tractor (22 litre/ha). The machines were tested in Malawi and Tanzania and produced average work rates of 30h/ha.

A computer program for the prediction of tractor work rate is introduced. It requires data on operating conditions such as field size, layout and soil strength, together with information on the tractor specification. Outputs provide predictions of work rate and cost per hectare. The program has been applied to a small farm situation in central Portugal and is currently being evaluated in Zimbabwe.

Rural Transport

A computer program to perform complex predictions of the performance and costs of vehicles used in rural transport operations has been developed. The program has been applied to transport situations in Malawi, Kenya, Portugal, Botswana and Ethiopia. The effect of operating conditions on various transport vehicles is described and it is concluded that small cheap vehicles may have a very high cost per tonne kilometre.

Dr Peter Crossley is a Lecturer in Engineering Design, Silsoe College, Silsoe, Bedford, England.

Child Health Care: Motivating Families and Communities to be More Involved and to Understand Measurement

DAVID MORLEY

My work has been to bring simple, curative, preventive and promotive health to children in developing countries. My particular concern has been to promote adequate growth of every child, as in this way undernutrition and malnutrition can be prevented. To achieve this, simple growth charts were introduced in the late 1950s and the use of these has spread to every developing country.

Only in the last decade has it become apparent (Figure 1) that nutrition and survival of small children depends more on whether their mother went to school than any other factor such as health services. In my retirement, it is my ambition to link these two together through bringing more understanding of measurements into families and communities. Three methods will be described and demonstrated: a direct recording scale for weighing children; a height measure with which children can measure each other's height accurately; lastly, a colour crystal thermometer which can replace glass and mercury thermometers for measuring body temperature.

Direct Recording Scale

This has been made feasible by developments in spring technology, so that springs can be produced which stretch 1 cm a kg to an accuracy of around one per cent. The child is placed in trousers or a sling attached to the lower end of the spring, which is then drawn up the chart by a cord attached to a beam in the house or the branch of a tree (Figure 2).

A plaster pointer cuffs the top of the spring and in this is a hole. The helper will fix this pointer and then the mother herself uses a ball pen through the hole to make the mark on her child's chart (Figure 3). For illiterate mothers, this may be

the first time they have used a ball pen. She will have seen the spring stretch with
the weight of her child and, with guidance, in time will come to understand the
meaning of a growth curve and how this depicts the change in weight and the
growth of her child. Previously, where readings were made from a dial and then

FIGURE 1

THE EDUCATION OF GIRLS IS CLOSELY ASSOCIATED WITH A FALLING
INFANT MORTALITY AND BIRTH RATE AND IMPROVED NUTRITION

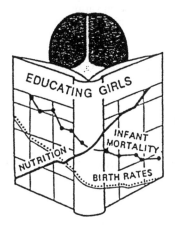

MOTHERS LITERACY, INFANT AND TODDLER MORTALITY, INDIA
Nut. News, Hyderabad, 9 No.2

	INFANT MORTALITY		TODDLER MORTALITY	
	Rural	Urban	Rural	Urban
Illiterate	132	81	29	17
Completed Primary	64	49	4	2

transferred to a chart, there have been opportunities for errors in charting and it
was difficult for the mother to comprehend what was happening.

Height Measure

This will depend on a plastic strip glued to the wall of the classroom. A very simple
bracket made in plastic will be available so that children can learn to measure each
other. Details of how to do this and how to fix the strip accurately will be printed
on the strip. Using the plastic bracket, the child will only be able to read one
number at a time (Figure 4) and in this way will get a direct reading in centimetres
and millimetres of the child being measured. It is hoped to produce this at
between US$5–10 and thus make it possible to have it widely available.

Colour Crystal Thermometer

Existing glass and mercury thermometers have a brief life in most developing
countries and are not suitable to provide to village health workers. The development of a colour crystal thermometer (Figure 5) has made it feasible to supply
these widely. They can be best used by being placed underneath the small child, as
the temperature gradient through a mattress is almost horizontal and, within ten
minutes, the temperature immediately under a child is almost identical with the
child's core temperature. These thermometers have been particularly designed to
recognise hypothermia, which is a problem, particularly in newborn and low birth

FIGURE 2

FIGURE 3

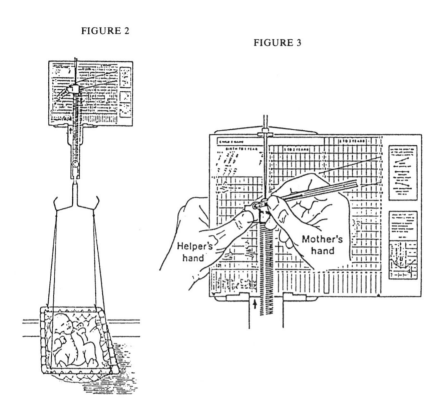

weight babies. The life of these thermometers is at least 14 years and they are almost indestructible; for example, they have been put through a domestic washing machine without any ill effect. However, if left for long periods in direct sunlight, the plastic will deteriorate. These thermometers cost around US$2 and may, in the future, empower the village health worker by making it possible for her to record children's temperatures.

FIGURE 4

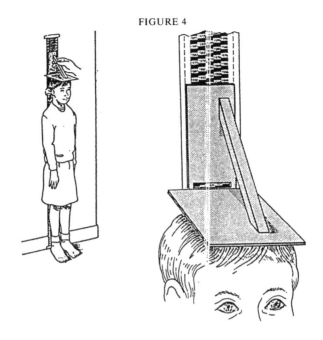

FIGURE 5

BODY TEMPERATURE WILL APPEAR IN GREEN ON THIS THERMOMETER

BROWN ....1°C higher than actual temperature reading

GREEN ....Correct body temperature reading

BLUE ....1°C lower than actual temperature reading

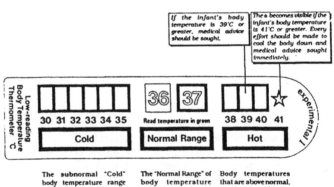

David Morley is Emeritus Professor of Tropical Child Health, University of London.

Strategies for Developing Food and Agro-Industries in Africa

J. V. S. JONES

It is argued that economic and industrial policies during the last quarter of a century have tended to make African countries more dependent on the industrial commodities or processed goods for which there is a limited demand. The limited number of modern industries that have been established are heavily dependent on imported inputs. When multinational firms are involved, opportunities are created for transferring profits (and hence potential investment resources) out of Africa which, together with the deteriorating terms of trade, slows down the rate of economic development. An important factor inhibiting the utilisation of local resources and local technology has been high exchange rate policies which make imported technology and materials cheaper than they otherwise would be.

Many large-scale food processing industries, often state sponsored, have accumulated large losses because of inadequate raw material supply. Some have been highly dependent on imported inputs, including spares. When foreign exchange has become scarce these have closed down. A case is made instead for developing a network of small-scale food industries which make better use of scarce capital resources and are less constrained by transport difficulties in raw material supply and underdeveloped marketing structures. Technology can be upgraded in stages as needed. Large-scale agro-industries may be more appropriate when raw material supply can be guaranteed and there are adequate markets. The textile industry is given as an example, and perhaps leather, which can produce intermediate materials.

The concept of 'integrated community development' is introduced as a means of mobilising the vast financial resources potentially available. It requires an ideological re-orientation on the part of professionals towards the realisation that ordinary people do have considerable expertise, and by blending this with modern scientific and technical knowledge, can give rise to more appropriate and more self-reliant solutions to the various problems of development. Education and training, correspondingly, needs to be re-orientated towards producing professionals more able to seek out local expertise and local sources of finance before turning to technological and financial inputs from outside, which, if on a recurrent basis, tend to perpetuate the debt crisis.

J. V. S. Jones is in the Department of Food and Nutritional Sciences, King's College, London.

Indigenous Technology and Industrialisation in Developing Countries: A Case Study of Nigeria

ADEREMI AJIBEWA

There was obviously indigenous technology in African society before the advent of colonialism. A country's technological competence cannot reliably be achieved by total dependence on others for its technology needs. Isolation from the rest of the world would of course retard progress in technology acquisition. However, it is the strengthening of the domestic scientific and technological base in developing countries which now demands urgent attention.

Nigeria's defunct 'Republic of Biafra' which was strangulated by the Federal Government during the Nigerian Civil War is cited as a case suggesting that isolation can have positive effects on the development of technology and notably on a society's ability to rely on its own inventive forces. One can say with some degree of certainty that Biafra Research and Production (RAP), in a mere fifteen months, fabricated and manufactured a wide range of weapons and kept many industrial establishments in operation. Biafran scientists and engineers demonstrated on a large scale the capability to refine petroleum and fuel oil in numerous and widely distributed locations without the assistance of expatriate technicians.

In this period of the Structural Adjustment Programme (SAP) firms have not started to look inwardly to substitute local raw materials for imported ones in the face of scarce and costly foreign exchange. The industrial research institutes (PRODA, FIIRO), realising that any significant effort towards self-reliance should begin with the production of basic consumer goods like food and clothing, have upgraded indigenous production systems through mechanisation of operations that were originally, traditionally done by hand – such innovations include completed plant for making *gari*, *fufu* production, the distilling of potable alcohol and mechanised lassara start production. Replacement, too, has taken place: composite flour for imported wheat for bread and bakery products; malted sorghum for imported barley; the production of table vinegar from palm wine and local fruits; soi-ogi production for maize; guinea corn (sorgum) and soya beans as substitutes for imported baby food (similac and lactogen).

Either by way of efficiency, quality of products, reduction in cost of production, use of local raw materials or adaptation to local environment, PRODA have introduced the washing machine, grinding machine, hammer mill, etc. However, insufficient attention to the selection of the environment clearly manifested in a lack of pre-project studies which have affected in no small measure the poor rate of adoption and diffusion of these innovations.

There is a tendency to raise the question of quality and precision as placing absolute limits on the technical feasibility of substituting local products for foreign importation. Initially these alternatives would not necessarily provide a superior product or a greater output than the multinationals provide and might be very costly, but whatever losses that might initially take place are part of the cost of improving national technology capability. They are likely to be short-term and would be compensated for by immediate social and political gains and in the long run by economic gains in terms of technological innovation and employment effects.

The argument therefore is to drive home the point that for a developing country like Nigeria what is needed is a determined and co-ordinated national effort and an ability to relate its R&D to its actual needs backed up by appropriate innovation policies. Nigeria would thus be able to improve its technological capability and develop feasible alternatives to some of the technologies imported into the country.

Aderemi Ajibewa is a Study Fellow, SPRU, University of Sussex, and also a staff member of the Department of International Relations, Obafemi Awolowo University, Nigeria.

Some Requirements for Building Effective Indigenous Technological Capabilities in Developing Countries

M. S. OUKIL

The paper deals with the issue of weaknesses of R&D systems in developing countries. It also aims at specifying the general and particular or minimum requirements for building effective indigenous technological capabilities in developing countries.

Following Professor Rosenberg's suggestion that during the 1990s the world economy will be shaped by the distribution of technological capabilities, the present paper centres around the following main questions: why are technological capabilities necessary in developing countries?; and, what are the minimum requirements for that purpose?

Certainly, the transfer of technologies from the North to the South has enabled the creation of production facilities, but that does not systematically solve technical problems. Thus, the development of technical infrastructure and the use of local potentialities are vital for recipient countries. In many of these, R&D systems exist but are generally weak and ineffective. One reason is that various necessary areas such as elements of organisation and management, linkages and interactions, and incentives are not properly set up.

The limited impact of science and technology policies and R&D programmes on productivity is largely due to the fact that very little consideration has been given to the micro-economic level where problems of production do arise. At the national level, according to Professor Freeman the problem is not of resources, but rather of their good organisation and management. In our view, and at both micro- and macro-economic levels, at least two requirements must exist in developing countries in order to develop reliable technological capabilities: targeting or responsiveness and seriousness. The latter refers to the manner in which technological innovations and R&D activities are undertaken and the way they are welcomed by the society in general.

M.S. Oukil is in the Department of Economics, University of Algiers.

Appendix: List of Papers Presented

Iftikhar Ahmad, *Scheme for the Establishment of New International Centres for Science and High Technologies*

A. Ajibewa, *Indigenous Technology and Industrialization in Developing Countries: A Case Study of Nigeria*

Helen Appleton and A. Jeans, *Technology from the People: Technology Transfer and Indigenous Knowledge*

P. Barker, R. Franceys and J. Pickford, *Environmental Upgrading for Low Income Communities in the South*

B. Cheze, *New Ways in Technological Research for Africa's Less Industrialised Tropical Countries*

N. Clark, *Development Planning and New Technologies*

A. Clunies Ross, *Internationalising the Costs of Environmental Measures*

P. Crossley, *Small-scale Engine-powered Mechanisation and Transport Devices for the Rural Sectors of Developing Countries*

J. Forje, *The State of Science and Technology in the Third World: The African Situation*

A. Haerian and H. Oraee, *Manpower Constraints in the Industrial Recovery of Post-War Iran*

A. Haerian and H. Oraee, *Industrial Renovation in Post-War Iran – The Need for Planning*

R. Heeks, *New Technology and the International Division of Labour: A Case Study of the Indian Software Industry*

Y. Hu, *China's Science and Technology – Transition and Dilemma*

M.M. Huq and K.M.N. Islam, *Transfer of Technology to Less Developed Countries: A Case Study of the Production of Fertiliser in Bangladesh*

J.V.S. Jones, *Strategies for Industrialization in Africa with Special Reference to the Food and Agro-Industry Sectors: Using Local Resources to Meet Africa's Needs*

K. King, *Science and Technology Manpower for the 1990s: The Role of the International Donor Community*

G. McRobie, *Technology Transfer from North to South*

E. Masanja, *Technology Transfer or Technology Development? – Third World Engineers' Dilemma*

D. Morley, *Motivating Families and Communities to be More Involved in and to Understand Measurement*

A. Narman, *Technical Manpower Development*

D. Newlands, *The Transfer of Medical Technology to Less Developed Countries*

S. Oukil, *Some Requirements for Building Effective Indigenous Capabilities in Developing Countries*

J. Pickett, *Indigenous Technological Capability in Sub-Saharan Africa*

B. Pollitt, *Mechanised Harvesting of Sugar Cane in Cuba*

K. Quasi and S Abdel Hamid, *Environmental Aspects of Technology Transfer to Developing Countries*

I. Raja and M. G. Dougar, *Solar Energy Research, Development and Demonstration in Pakistan*

F. O'Reilly, *Education of Farmers for Technology Transfer in the Third World*

T. Thomas, *University Research in Britain for Rural Development in Africa: The Example of the Hydraulic Ram Pump*

M. Tribe, *Appraising Small-Scale Modern Dairy Developments in Kenya – Opportunities and Pitfalls*

J. Twidell, *The Role of the Energy Studies Unit in Technology Transfer*

G. Wapler, *Transfer of Applied Technology*

A. A. Williams, N. P. A. Smith and S. Mathema, *Application of Appropriate Technology to Small-Scale Hydro-Electric Power*

A. Young, *The Semiconductor Industry: Its Contribution to Third World Development*